SpringerBriefs in Molecular Science

Chemistry of Foods

Series Editor

Salvatore Parisi, Al-Balqa Applied University, Al-Salt, Jordan

The series Springer Briefs in Molecular Science: Chemistry of Foods presents compact topical volumes in the area of food chemistry. The series has a clear focus on the chemistry and chemical aspects of foods, topics such as the physics or biology of foods are not part of its scope. The Briefs volumes in the series aim at presenting chemical background information or an introduction and clear-cut overview on the chemistry related to specific topics in this area. Typical topics thus include:

– Compound classes in foods—their chemistry and properties with respect to the foods (e.g. sugars, proteins, fats, minerals, …)
– Contaminants and additives in foods—their chemistry and chemical transformations
– Chemical analysis and monitoring of foods
– Chemical transformations in foods, evolution and alterations of chemicals in foods, interactions between food and its packaging materials, chemical aspects of the food production processes
– Chemistry and the food industry—from safety protocols to modern food production

The treated subjects will particularly appeal to professionals and researchers concerned with food chemistry. Many volume topics address professionals and current problems in the food industry, but will also be interesting for readers generally concerned with the chemistry of foods. With the unique format and character of SpringerBriefs (50 to 125 pages), the volumes are compact and easily digestible. Briefs allow authors to present their ideas and readers to absorb them with minimal time investment. Briefs will be published as part of Springer's eBook collection, with millions of users worldwide. In addition, Briefs will be available for individual print and electronic purchase. Briefs are characterized by fast, global electronic dissemination, standard publishing contracts, easy-to-use manuscript preparation and formatting guidelines, and expedited production schedules.

Both solicited and unsolicited manuscripts focusing on food chemistry are considered for publication in this series. Submitted manuscripts will be reviewed and decided by the series editor, Prof. Dr. Salvatore Parisi.

To submit a proposal or request further information, please contact Dr. Sofia Costa, Publishing Editor, via sofia.costa@springer.com or Prof. Dr. Salvatore Parisi, Book Series Editor, via drparisi@inwind.it or drsalparisi5@gmail.com

More information about this subseries at http://www.springer.com/series/11853

Michele Barone · Alessandra Pellerito

Sicilian Street Foods and Chemistry

The Palermo Case Study

 Springer

Michele Barone
Associazione 'Componiamo
il Futuro' (CO.I.F.)
Palermo, Italy

Alessandra Pellerito
Food Safety Consultant
Palermo, Italy

ISSN 2191-5407 ISSN 2191-5415 (electronic)
SpringerBriefs in Molecular Science
ISSN 2199-689X ISSN 2199-7209 (electronic)
Chemistry of Foods
ISBN 978-3-030-55735-5 ISBN 978-3-030-55736-2 (eBook)
https://doi.org/10.1007/978-3-030-55736-2

This Springer imprint is published by the registered company Springer Nature Switzerland AG
The registered company address is: Gewerbestrasse 11, 6330 Cham, Switzerland

Contents

About the Authors

Michele Barone is an experienced consultant working in the field of food science and technology, and also in restoration chemistry. His work in food science focuses mainly on food packaging and correlated failures, and selected food products with a dedicated tradition (for instance, the Mediterranean Diet). More recently, he has written about food traceability systems in the field of European cheese products. Michele currently works at the Association 'Componiamo il Futuro' (CO.I.F.) in Palermo, Italy (sector: professional training). He has recently published *Chemicals in the Food Industry*, *Dietary Patterns, Food Chemistry and Human Health*, and *Quality Systems in the Food Industry* in the SpringerBriefs in Chemistry of Foods.

Alessandra Pellerito is a Biologist graduated at the University of Bologna, Italy (2013) with full marks (110/100 cum laude) after the initial B.Sc. in Biology (Palermo, Italy). After a short period spent in the UK, Dr. Pellerito moved to Germany (Magdeburg), where she has been working for 2 years as Research Assistant at the Leibniz Institute for Neurobiology. At present, Dr. Pellerito works as a Food Consultant in the private sector. Alessandra's first article on food chemistry—Mania I, Barone C, Pellerito A, Laganà P, Parisi S, *Trasparenza e Valorizzazione delle Produzioni Alimentari. L'etichettatura e la Tracciabilità di Filiera come Strumenti di Tutela delle Produzioni Alimentari*. Industrie Alimentari—concerns authenticity problems, traceability, and food labelling in the current European market. She has also co-authored *Antimicrobial Substances for Food Packaging Products: The Current Situation, Journal of AOAC International*, in 2018.

Chapter 1
The Street Food Culture in Europe

Abstract At present, the food market shows apparently the prevailing position of fast-food operators in many Western Countries. However, the concomitant presence of the traditional lifestyle model known as the 'Mediterranean Diet' in Europe should not be excluded. The fragmentation of the current market of foods and beverages––sometimes discussed in terms of 'Food Wars'—involves many possible products and lifestyles, including vegan foods, religion-based diets, 'healthy' or fast-weight loss diets, organic foods, Mediterranean Diet… and also the so-called 'Street Foods'. These products are diffused as a cultural heritage in all known urbanised areas of the world, suggesting relationships with social aggregation, economic convenience, typical folk elements, etc. The study of street foods can help when speaking of the examination of several historical products of Sicily, Italy, and particularly of the largest Sicilian city, Palermo.

Keywords Cultural heritage · Fast food · HACCP · Junk food · Mediterranean diet · Street food · Vegan food

Abbreviations

AFSUN	African Food Security Urban Network
BBC	British Broadcasting Corporation
FAO	Food Agriculture Organization of the United Nations
FBO	Food businesssss operator
HACCP	Hazard Analysis and Critical Control Points
INFOSAN	International Food Safety Authorities Network
MD	Mediterranean Diet
MPRA	Munich Personal RePEc Archive
SARS-COV-2	Severe acute respiratory syndrome-associated coronavirus
SARS	Severe acute respiratory syndrome
SF	Street food
WHO	World Health Organization

1.1   An Introduction to 'Food Wars'

At present, the food market shows apparently the prevailing position of fast-food operators in many Western Countries. It should be recognised that the biggest industrial companies in the ambit of foods and beverages promote their fast-consumption products with the aim of achieving more and more portions of the whole food market everywhere. We are considering now the broad ambit of marketing strategies, because there are not success stories without good or excellent advertising techniques (Buijzen et al. 2008; Council on Communications and Media and Strasburger 2011; Dixon et al. 2007; Lee et al. 2009; Zimmerman and Bell 2010). It has to be also considered that the modification of food eating styles and behaviours cannot be possible without a profound modification of the whole food chain system on a large scale, in terms of food globalisation (Menezes 2001; Oosterveer and Sonnenfeld 2012; Raynolds 2004; Swinnen and Maertens 2007) and technological choices (Scholliers and Van Molle 2005). In other terms:

(a) The modification of production processes in some steps of the food chain can turn the whole process and the consequent food product (and related service) into a really different version of the initial idea of food (or beverage product)

(b) The more the number of modified steps, the higher the difference between the initial (historical, traditional, currently accepted) idea of food or beverage product and the final result

(c) At the same time, the behaviour and/or consumeristic trends of normal food consumers (Parisi 2012, 2013) can evolve towards a new or 'improved' definition of trending behaviour, with implicit risks in terms of loss of historical traditions

(d) Moreover, the continuous improvement of new recipes and foods/beverages can partially or totally remove the remembering of 'ancient' foods from the collective memory of food consumers, both on the national/regional/local level and in a large-scale ambit

(e) The possible end of such a story may be the complete loss of information concerning old foods and beverages, taking also into account that lost/erased/removed information may concern both literature traditions and traditional/technological techniques at the same time.

In addition, and it is not a marginalised aspect of the whole matter concerning foods and beverages, these products are often associated with wars in the real meaning of this term. Food can be considered as a weapon obtained by means of production systems that may be destroyed during real conflicts. Moreover, the lack (or the possible lack in future) of food commodities can spread across the world with the observable 'food insecurity' phenomenon (Messer 1994, 1996; Messer and Cohen 2001, 2007; Messer and Uvin 1996; WHO 2020). Recently, the new severe acute respiratory syndrome (SARS)-associated coronavirus (SARS-COV-2) (Hoehl et al. 2020) pandemic has forced food producers, food banks, and many productive/distributive systems to face an enormous and large-scale demand for food supplied in many countries with consequent panic everywhere (Cachero 2020;

Government of Canada 2020; Hayes 2020; Martin 2020; Shaw 2020). The role of food production systems is now critical both in real famine scenarios and in non-critical ambits when speaking of food supplies. The globalisation has naturally amplified the action of similar emergencies (Andrée et al. 2014; Lang 1999; Lappé et al. 1998).

1.2 Cultural Heritage and Consumeristic Behaviours

The preference for new/improved foods and beverages with a meaning centred mainly on the achievement of main market portions is a practical risk when speaking of cultural heritage. After all, cultural traditions define all human beings, Nation by Nation and area by area (Alfiero et al. 2017; Delgado et al. 2017; Maranzano 2014; Wilkinson 2004).

On the other hand, the current food market—in industrialised Countries at least— is not only a matter of industrial and fast-consumption foods. It has to be recognised that the ambit of food productions and the consequent behaviour of food consumers is always fluctuating between two opposite extremities (Almli et al. 2011; Blair 1999; FAO 1997; Khairuzzaman et al. 2014; Sezgin and Sanlier 2016; Sperling 2010; Steyn and Labadarios 2011; Steyn et al. 2013; Verbeke and Viaene 2000):

(1) The preference for highly palatable, soft, and 'delicious' products with (unfortunately) a relevant amount of chemical compounds related to notable caloric intakes, on the one hand, and
(2) The difficult search for 'natural', 'historical', 'traditional', 'healthy', and/or ethically approvable foods and beverages on the other hand. The group of vegan and vegetarian foods and beverages, being probably associated with some 'vegan' religion, has to be considered in this ambit (BBC 2020).

The contrast between these two different consumeristic behaviours should be considered with care because of the coexistence of different trends in the current world of food and beverage consumers. The simple example of vegetarian foods should be clear enough because of the relationship between a limited group of well-being consumers on the one hand and a little portion of the whole market of foods and beverages. In addition, the simple group of vegetarian-style diets can be further subdivided into four sub-types at least:

(a) Vegan style (Batte et al. 2007; Corvo 2016; Dedehayir et al. 2017; Marcus 2000; Seyfang 2009; Souza et al. 2020; Weckroth 2018),
(b) Lacto-ovo vegetarian style (only milk, cheese, yogurt, eggs, and vegetable foods),
(c) Lacto-vegetarian style (only milk, cheese, yogurt, and vegetable foods),
(d) Ovo vegetarian style (milk, cheese, and yogurt; eggs can be accepted).

The fragmentation of such a niche market should explain very well the nature of the current situation when speaking of foods and beverages today. It has to be considered that the entire group of modern diet styles may be remarkable (Fig. 1.1):

Fig. 1.1 Some of the most notable diet styles in the present age

(1) Religion-based diets: macrobiotics or Buddhist diets, Hindu, Halal, Kosher, etc.,
(2) 'Healthy' diets: low-calorie, very low-calorie, low-carbohydrate, low-fat…,
(3) Fast-weight loss diets,
(4) 'Detox' diets (elimination of colourants and preservatives; consumption of dietary supplements; assumption of water in notable amounts), Example: activated charcoal diet,
(5) Food-specific dietary advices (consumption of one food type or category only),
(6) Vegetarian diets,
(7) Organic-food based diets.

In this ambit, the Mediterranean Diet (MD) cannot be excluded (Delgado et al. 2017), similarly to other historical/traditional diet styles. It has to be considered that this choice is a heritage of past cultures, while the above-mentioned styles are modern trends, generally oriented to non-historical or traditional reasons, and sometimes questionable or unreliable, depending on types and related (claimed) health effects.

1.3 Food Products and Socio-economic Importance in the Current World

Another interpretation of the current market of foods and beverages—by the view-point of the consumer—can be the socio-economic importance (or value) of selected edible products. In other words, the production and the distribution (with non-productive activities such as marketing) of selected foods or beverages with peculiar attributions can (Alves da Silva et al. 2014; FAO 2009; Hopping et al. 2010; Long-Solís 2007; Maranzano 2014; Privitera 2015; Steyn and Labadarios 2011; Steyn et al. 2013; Tinker 1999):

(1) Promote the employment,
(2) Enhance tourism in certain Countries,
(3) Defend cultural heritage, as above mentioned,
(4) Give a defined nutritional value to consumer.

The opposite face of this reflection may be correlated with the economic value of foods and beverage when consumers perceive that the required prices are excessive. In other words, there is a certain association between the price of foods and beverages and a determined dietary style. Two examples can be examined with relation to the current market (Carfì et al. 2018a, b; Hughner et al. 2007; Lukić 2011; Marangon et al. 2016; Nasir and Karakaya 2014; Yiridoe et al. 2005):

(a) Organic foods tend to be sold at higher prices if compared with 'conventional' products. The reason is linked to many aspects, including also the search for fresh (unprocessed or minimally processed) foods, 'natural' tastes, excellent good-looking, seedless, etc. The production of such a food (similar or perceived equal to the 'natural version' of the idea of food product) should explain and justify the higher price of organic foods, and concerned consumers are ready to pay for this type of foods

(b) Vegetarian products, including vegan foods, show a similar trend because of the perceived value of vegetarian consumers. In brief, the search for foods without animal raw materials can justify the higher price for similar food and beverage products, and concerned consumers are ready to pay for this type of foods. Moreover, conventional food producers and vegan-food producers may establish commercial alliances or cooperative agreements with the aim of enhancing respective market shares instead of empowering commercial contrasts (also named 'food wars').

On these bases, the relationship between a determined food product—inked with a specified dietary style—and its economic price can be evaluated correctly. As a consequence, it may be assumed that low prices are correlated with conventional, large-consumption, and high-marketed products, while the opposite relationship (high prices = non-conventional foods) means tacitly that related foods and beverages are part of a market 'niche', depending on the peculiar lifestyle. This assumption should be always taken into account.

1.4   Junk Foods Versus Conventional Foods or Fast-Food Services Versus Mediterranean Diet?

A correlated addition to this reflection would concern the matter of 'healthy' and 'conventional' foods on the one side as opposed to so-called 'junk foods' (Chapman and Maclean 1993; Datar and Nicosia 2012; Yaniv et al. 2009; Sacks et al. 2011; Wiles et al. 2009). The last words explicitly mean that related products are not healthy enough because of the excessive amount of fat matters and sugars, with consequent negative effects on human health (obesity, diabetes, etc.).

In this heterogeneous ambit, the presence of the traditional lifestyle model known as the 'Mediterranean Diet' in Europe cannot be excluded. In opposition to the fast-food culture, the Mediterranean Diet pattern has been increasingly adopted in many Countries, on the basis of reported studies concerning human health and quality of life (Curtis and O'Keefe 2002; De Batlle et al. 2008; Delgado et al. 2017; Mone and Bulo 2012; Shadman et al. 2014). On the other side, some unexpected effects of the counterposition between Mediterranean-Diet and fast-food styles have been observed in the last decades, including the proposal of Mediterranean-Diet-like foods for out-of-home consumption.

1.5   Mediterranean Foods or Street Foods? An Analysis

The proposed strategy (Mediterranean-Diet-like foods for out-of-home consumption) is a commercial operation concerning also the so-called 'Street Food', reported to be a different version of Mediterranean foods (in the Mediterranean Basin, including North Africa and Middle East regions) (Chammem et al. 2018; David 2002; Delgado et al. 2017; Haddad et al. 2020a, b; Helou 2006; Matalas et al. 2001). However, the relationship between the Mediterranean Diet and Street Foods is not exact and geographically correct.

With exclusive relation to Mediterranean dietary styles, it has to be recognised that related consumers prefer mainly vegetable products and transformed foods from vegetable sources, while the consumption of dairy and meat foods is low (Keys 1995). This consideration is the basis for the famous Mediterranean Diet pyramid (Bach-Faig et al. 2011; Davis et al. 2015; Willett et al. 1995), while selected foods and beverages are placed, level by level, with relation to a recommended amount (or a number of times) per week (Delgado et al. 2017).

Anyway, because of the important contribution of such an eating style guide to wellness in human beings as generally observed in certain Mediterranean areas such as Greece, Spain, Portugal, Tunisia, and Southern Italy, Mediterranean Diet is highly recommended as a good preventive 'therapy' against cardiovascular diseases, diabetes, and other modern diseases. Consequently, MD is perceived and recommended—also commercially—as a synonym of 'safe', 'healthy', and 'hygienic' food, even if recommendations do not refer to chemical, physical, and microbiological food risks in the ambit of 'Hazard Analysis and Critical Control Points' (HACCP) evaluations. In fact, advantages of MD-foods are mainly related to their composition (low calories; low carbohydrates, sugars, and lipids), and the presence of antioxidants including polyphenols (Haddad et al. 2020a, b; Parisi 2019, 2020).

On the other side, 'street food' (SF) is not a synonym of MD foods. On the contrary, the definition of SF—according to the Food Agriculture Organization of the United Nations (FAO)—concerns only ready-to-eat foods and beverages which are prepared and sold by different food business operators (FBO) without an immobile location. In other terms, these foods and beverages are literally produced and sold in the streets and in similar places, including also small food trucks (Anenberg and Kung 2015; FAO 1997; Pitte 1997; Vanschaik and Tuttle 2014).

On these bases, it should be recognised that SF is not a peculiar feature of MD-related areas. On the contrary, many examples have been always observed worldwide:

(1) The European Union and single National Countries in this area, since ancient Greece and the Roman Empire, including also the old Ottoman Empire (Heinzelmann 2014; Larcher and Camerer 2015; Maranzano 2014; McHugh 2015; Simopoulus and Bhat 2000; Wilkins and Hill 2009).

(2) African Countries such as Nations in the *Maghreb* area, at present and since Ancient Egypt at least (Larcher and Camerer 2015; Sheen 2010).

(3) North- and South American Countries and cultures. Examples: Mexico, Belize, Bolivia, Brazil, Colombia, and the Caribbean Region (Cortese et al. 2016; de Suremain 2016; Harris 2003; Kollnig 2020; Morton 2014; Pilcher 2017; Webba and Hyatt 1988).

(4) Asian Regions, with a peculiar reference to historical traditions of certain Nations such as Imperial China and Singapore where the term 'food alley' is preferred (Bell and Loukaitou-Sideris 2014; Chong and Eun 1992; Manning 2009; Steven 2018). On the other side, areas such as India appear to know SF only in recent times (Olsen et al. 2000; Patel et al. 2014). Thailand, Singapore, and Vietnam should be considered at present.

(5) Oceanian Nations such as Australia (Booth and Coveney 2007).

Anyway, SF is diffused as a cultural heritage in all known urbanised areas of the world, suggesting a clear correlation between social aggregation and food consumption. By this viewpoint, SF is not so different from foods served at fast-food restaurants and so on, with the particular difference that there are not sold into immobile locations. The correlation with urbanisation is substantially sure and well accepted, above all in developing Countries where a notable aggregation of different people has been progressively observed in the past century and in a few limited cities. The phenomenon is particularly evident in several African urban and sub-urban areas and in India also, while historical centres with more than 2000 years of history propose similar situations with the same basic features: lack of immobile locations, economic convenience, 'poor food' characteristics, typical folk elements, and a certain trend towards poor hygienic conditions (Adjrah et al. 2013; Chukuezi 2010; Crush et al. 2011; Hawkes et al. 2017; Sujatha et al. 1997).

1.6   Street Food or Poor Foods?

Another important—and probably critical characteristic—of SF is linked to the nature of served foods as linked with economic convenience. It should be admitted that one—or the most important—of the key points for the enormous success of SF is their low cost (Fellows and Hilmi 2012), tacitly meaning that this category of served foods is also preferred by 2.5 billions of consumers worldwide because of its historical origin: 'poor food' (Chukuezi 2010).

These words mean implicitly that SF is or can be preferred because of two distinct and possibly contrasting elements (Allison 2018; Dittrich 2017; Lee 2017; Rosales 2014):

(1) SF can supply a good nutritional contribution to people, especially in disadvantaged areas, because of the notable amount of bioavailable proteins and energy intakes. This situation can be easily observed because SF is the heritage of centuries of historical traditional cooking, and ancient foods tended to supply iron, energy, and proteins in remarkable amounts

(2) On the other hand, SF is cheap enough if compared with other food choices. The correlation between price and food is tacit: the higher the price, the higher the inherent value of the food in terms of raw materials, hygienic preparation, and so on. However, the advantage for consumers (in terms of price) may be perceived at the same time as a negative factor because 'poor' (inadequate, hygienically poor, etc.) food could be sold at high prices. Years ago, the use of SF might have been observed and perceived by consumers as a sign of social discrimination.

At present, the situation observed worldwide has changed visibly. The contrast between the main type of food consumers searching for tasty, cheap, and fast-food products on the one hand, and the increasing group of different food consumers aspiring to find traditional, hygienically safe, minimally processed, and ethically acceptable products on the other hand, has caused the fragmentation of the current food and beverage market. In addition, there is the possibility that the search for traditional foods such as real '*Italian pizza*' or other foods (*Kebab*, etc.) may direct several consumers towards traditional and hygienically questionable foods at the same time. This situation is currently observed when speaking of SF in certain areas. The World Health Organization (WHO) has already defined its 'Five Keys to Safer Food' with the aim of giving some correct guidelines to be followed also in the SF ambit (INFOSAN 2010). These rules (WHO Department of Food Safety, Zoonoses and Foodborne Diseases 2006) can be expressed as follows (Fig. 1.2):

(a) Sanitisation,
(b) Separation between raw materials and cooked foods,
(c) High cooking temperatures,
(d) Use of safe temperatures (refrigerated or frozen storage, or room temperature depending on situations),
(e) Use of safe water and raw ingredients/additives.

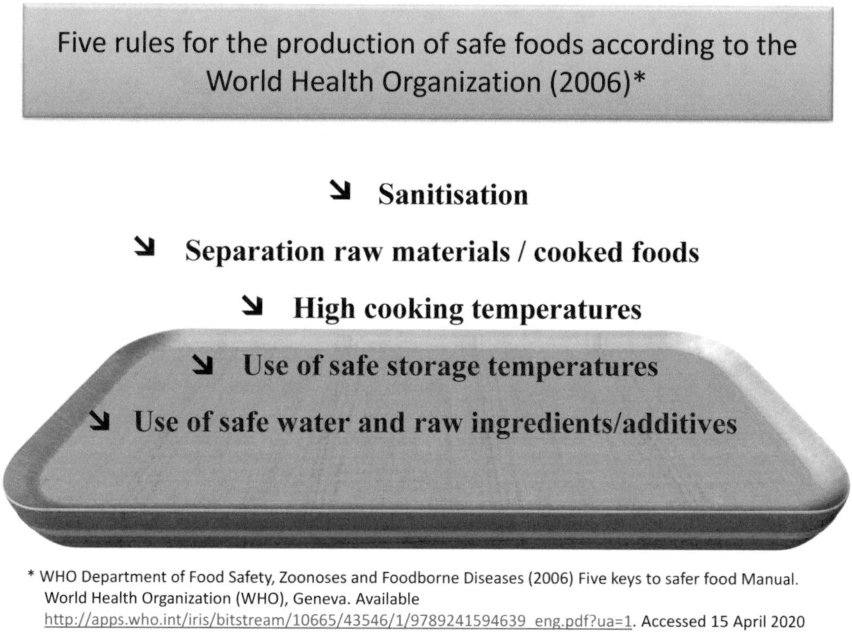

Five rules for the production of safe foods according to the
World Health Organization (2006)*

↘ **Sanitisation**

↘   **Separation raw materials / cooked foods**

↘   **High cooking temperatures**

↘   **Use of safe storage temperatures**

↘ **Use of safe water and raw ingredients/additives**

* WHO Department of Food Safety, Zoonoses and Foodborne Diseases (2006) Five keys to safer food Manual.
World Health Organization (WHO), Geneva. Available
http://apps.who.int/iris/bitstream/10665/43546/1/9789241594639_eng.pdf?ua=1. Accessed 15 April 2020

Fig. 1.2   Five rules for the production of safe foods according to the World Health Organization
(2006)

However, these rules are not always respected in the SF ambit. In detail, it may be
observed that SF vendors generally use poor (inadequate) ingredients; they work into
hygienically unacceptable or poor locations, using a few personal hygiene procedures
only. Naturally, the lack of controls and official inspections does not favour the use of
recommended guidelines according to different authors (Adjrah et al. 2013; Azanza
et al. 2000; De Souza et al. 2015; Liu et al. 2014; Mchiza et al. 2014; Omemu and
Aderoju 2008; Sabbithi et al. 2017; Thakur et al. 2013).

At present, more and more vendors worldwide have to maintain their traditions
with an eye on food safety and hygiene. The 'take away' services in New York
City of food truck services in Italy are a good demonstration of this multi-cultural
phenomenon: after all, there are urban areas such as Shangai (China) where 90%
of the global food consumption may be found in the SF ambit, and consequently a
minimum set of hygienic rules have to be requested and demonstrated (Maranzano
2014).

In this ambit, the study of SF in the world should be promising enough. This
book is dedicated to the study of a peculiar type of folk cuisine found in Sicily, Italy,
and particularly in the largest city, Palermo. However, a simple description of SF in
Europe could be useful enough before describing Sicilian and Palermo's SF: in fact,
these products are the cultural heritage not only of a specific area in the Mediterranean
Basin, but also of different and sequentially chronological civilisations found in Sicily

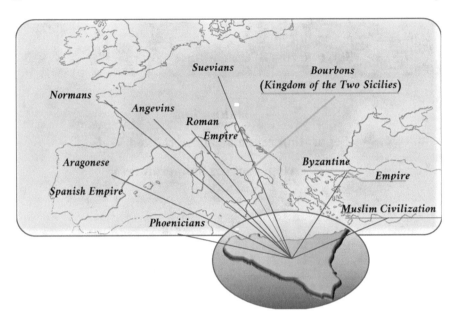

Fig. 1.3 The Mediterranean history in a single island: Sicily. Phoenician, Roman, Byzantine, Muslim, Norman, Suevian, Angevin, Aragonese, Spanish, and the Bourbon periods have to be mentioned because of their influence on Sicilian cuisine and SF products. This image shows a historical map identifying the peculiar civilisations with their original (and possibly) current localisation in the Mediterranean Basin

(Fig. 1.3). A brief overview of currently existing SF specialities in Europe should be given, even if the description is surely incomplete.

1.7   Street Food in Europe. New Products and Old Civilisations

Street food is a matter of history, cultural heritage, business and advertising strategies, and food chemistry/microbiology/technology at the same time. It could be very difficult to exhibit all the possible SF in current Europe and the broader area of the Mediterranean Basin (including the *Maghreb* and Middle East Countries). For this reason, this Section will briefly give an overview of the most famous and/or culturally important SF widely known and recognised in the European Countries (Kraig and Sen 2013).

Table 1.1 shows a list of European Nations where one SF at least can be found and currently purchased. It should be considered that each mentioned SF has possibly one or more geographical areas into the same Nation with some difference. The description offered in Table 1.1 cannot be exhaustive; on the other side, it should

Table 1.1 A list of European Countries with at least one currently available SF (Kraig and Sen 2013)

Country	SF type (National or typical name)	Country	SF type (National name)
Austria (Heinzelmann 2014; Kraig and Sen 2013; O'Mahony 2004; Raak et al. 2020; Rohm 1990)	*Kaiserschmarnn* *Käsekrainer* *Bratwurst* *Frankfurter* *Currywurst*	Ireland (Nicolay et al. 2011)	Chicken and stuffing sandwiches Yellow Man Irish Toffee
The Balkans Area (Mlekuz 2015)	*Burek*		
Belarus(Romero-Gwynn et al. 1997)	*Nalistniki*	Italy, excluding Palermo city (Amabile 2004; D'Acampo 2015; Gambella et al. 2008; Giammanco et al. 2011; Parente 2007)	Focaccia of Recco *Frisceu* of Genoa *Malloreddus* *Sebadas* *Culurgione* *Pizzi* of Lecce Sicilian "primosale" *Pecorino* cheese
Belgium (Stengs 2013)	*Frieten met Mayonnaise*	The Netherlands (De Graaf 2005)	*Oliebollen*
Denmark (Eames et al. 2016; Sørensen et al. 2018)	*Æbleskiver* Pork loin sandwich with sweet-and-sour red cabbage Pork loin with crisp crackling	Norway (Petrenya et al. (2019)	*Fiskekaker* *Lomper*
France (Bakker and Clarke 2011; Liakopoulos et al. 2015; Puckette and Hammack 2015)	Crêpes Traditional hot wine with Alsace	Poland (Kowalczyk and Korpysz 2020)	*Kremówka Papiesk* *Zapiekanka*
Germany	*Buletten* or *Frikadellen*	Romania (Uğurlu 2015)	*Mititei/Mici*
United Kingdom (Olsen et al. 2000)	Fish & Chips	Russian Federation (Breiter 1997; Maiatsky 2009)	*Pirozhki* *Shashlyk*
Greece (Soldatou et al. 2009; Vareltzis et al. 2016)	*Souvlaki* κολλύριο (*kollyrio*, a typical small round bread) *Tzatziki* sauce	Turkey (Erkaya et al. 2015)	*Izgara* *Kimyonlu Köfte* *Nohutlu Pilav*

These SF are often an expression of one or more different locations into the same Country with some difference, and generally advertised by means of dedicated SF Festivals demonstrate the peculiar brilliance of the European tradition when speaking of food and beverage heritages. It has to be also noted that many of these products tend to show certain similarities, such as the situation of

(a) The Russian-style *Pirozhki* dish associated with a typical *Bolognese* (Italian-style) sauce (Maiatsky 2009).
(b) The influence of Balkan traditions on the current recipe of Romanian-style *mititei* or *mici* (Uğurlu 2015).
(c) And, last but not least, the '*pizza*' and '*kebab*' invasion in many European cultures with the possible results of the Austrian *kebab* version (*dörum*) or the Balkan-style food named '*pizza/burek*' (Brown 2018; Silverman 2012).

In brief, SF is one of the most interesting European 'souls' when speaking of social food out-eating. Because of the importance of this phenomenon in the history and in European Countries at least (with a significant invasion of related traditions in the United States of America at least), the SF 'explosion' should be considered. The so-called 'Palermo case study' can be extremely useful in this ambit.

References

Adjrah Y, Soncy K, Anani K, Blewussi K, Karou DS, Ameyapoh Y, de Souza C, Gbeassor M (2013) Socio-economic profile of street food vendors and microbiological quality of ready-to-eat salads in Lomé. Int Food Res J 20(1):65–70

Alfiero S, Giudice AL, Bonadonna A (2017) Street food and innovation: the food truck phenomenon. Brit Food J 119(11):2462–2476. https://doi.org/10.1108/BFJ-03-2017-0179

Allison N (2018) Food, the city, and the street. Rev Adm Empresas 58(3):345

Almli VL, Verbeke W, Vanhonacker F, Næs T, Hersleth M (2011) General image and attribute perceptions of traditional food in six European countries. Food Qual Preference 22(1):129–138. https://doi.org/10.1016/j.foodqual.2010.08.008

Alves da Silva S, Cardoso RCV, Góes JTW, Santos JN, Ramos FP, Bispo de Jesus R, Sabá do Vale R, Teles da Silva PS (2014) Street food on the coast of Salvador, Bahia, Brazil: a study from the socioeconomic and food safety perspectives. Food Control 40(1):78–84. https://doi.org/10.1016/j.foodcont.2013.11.022

Amabile F (2004) Mangiare per strada. Airplane, Bologna

Andrée P, Ayres J, Bosia M, Mássicotte MJ (eds) (2014) Globalization and food sovereignty: global and local change in the new politics of food. University of Toronto Press, Toronto

Anenberg E, Kung E (2015) Information technology and product variety in the city: the case of food trucks. J Urban Econ 90:60–78. https://doi.org/10.1016/j.jue.2015.09.006

Azanza MPV, Gatchalian CF, Ortega MP (2000) Food safety knowledge and practices of street food vendors in a Philippines University campus. Int J Food Sci Nutr 51(4):235–246. https://doi.org/10.1080/09637480050077121

Bach-Faig A, Berry EM, Lairon D, Reguant J, Trichopoulou A, Dernini S, Medina FX, Battino M, Belahsen R, Miranda G, Serra-Majem L (2011) Mediterranean diet pyramid today. Science and cultural updates. Pub Health Nutr 14(12A):2274–2284. https://doi.org/10.1017/S136898001100 2515

Bakker J, Clarke RJ (2011) Wine: flavour chemistry. Wiley, Hoboken

Batte MT, Hooker NH, Haab TC, Beaverson J (2007) Putting their money where their mouths are: Consumer willingness to pay for multi-ingredient, processed organic food products. Food Pol 32(2):145–159. https://doi.org/10.1016/j.foodpol.2006.05.003

BBC (2020) Ethical veganism is philosophical belief, tribunal rules. British Broadcasting Corporation (BBC), London. Available https://www.bbc.com/news/uk-50981359. Accessed 14 April 2020

Bell JS, Loukaitou-Sideris A (2014) Sidewalk informality: an examination of street vending regulation in China. Int Plan Stud 19(3–4):221–243. https://doi.org/10.1080/13563475.2014.880333

Blair D (1999) Street food vending and nutritional impact. Agric HumValues 16(3):321–323

Booth SL, Coveney J (2007) Survival on the streets: prosocial and moral behaviors among food insecure homeless youth in Adelaide, South Australia. J Hunger Environ Nutr 2(1):41–53. https://doi.org/10.1080/19320240802080874

Breiter M (1997) What is the difference between "shashlyk" and "barbecue"? Perspect Stud Translatol 5(1):85–100. https://doi.org/10.1080/0907676X.1997.9961302

Brown K (2018) Burek, Da! Sociality, context, and dialect in Macedonia and beyond. In: Montgomery DW (ed) Everyday life in the Balkans. Indiana University Press, Bloomington

Buijzen M, Schuurman J, Bomhof E (2008) Associations between children's television advertising exposure and their food consumption patterns: a household diary–survey study. Appet 50(2–3):231–239. https://doi.org/10.1016/j.appet.2007.07.006

Cachero P (2020) Wuhan residents on coronavirus lockdown are facing food shortages. www.businessinsider.com. Available https://www.businessinsider.com/wuhan-residents-coronavirus-lockdown-forced-order-food-apps-delivered-home-2020-3?IR=T. Accessed 14 April 2020

Carfì D, Donato A, Panuccio D (2018a) A game theory coopetitive perspective for sustainability of global feeding: agreements among vegan and non-vegan food firms. In: Khosrow-Pour M (ed) Game theory: breakthroughs in research and practice. IGI Global, Hershey, pp 71–104

Carfì D, Donato A, Schilirò D (2018b) Sustainability of global feeding. Coopetitive interaction among vegan and non-vegan food firms. Munich Personal RePEc Archive (MPRA), the Munich University Library, Munich. Available https://mpra.ub.uni-muenchen.de/88400/. Accessed 14 April 2020

Chammem N, Issaoui M, De Almeida AID, Delgado AM (2018) Food crises and food safety incidents in European Union, United States, and Maghreb Area: current Risk communication strategies and new approaches. J AOAC Int 101(4):923–938. https://doi.org/10.5740/jaoacint.17-0446

Chapman G, Maclean H (1993) "Junk food" and "healthy food": meanings of food in adolescent women's culture. J Nutr Educ 25(3):108–113. https://doi.org/10.1016/S0022-3182(12)80566-8

Chong HK, Eun VLN (1992) 4 Backlanes as contested regions: construction and control of physical. In: Huat CB, Edwards N (eds) Public space: design, use and management. Singapore University Press, Singapore

Chukuezi CO (2010) Food safety and hyienic practices of street food vendors in Owerri, Nigeria. Stud Sociol Sci 1(1):50–57. https://doi.org/10.3968/j.sss.1923018420100101.005

Cortese RDM, Veiros MB, Feldman C, Cavalli SB (2016) Food safety and hygiene practices of vendors during the chain of street food production in Florianopolis, Brazil: a cross-sectional study. Food Control 62:178–186. https://doi.org/10.1016/j.foodcont.2015.10.027

Corvo P (2016) Food culture, consumption and society. Palgrave Macmillan, London. https://doi.org/10.1057/9781137398178

Council on Communications and Media, Strasburger VC (2011) Children, adolescents, obesity, and the media. Pediatr 128(1):201–208. https://doi.org/10.1542/peds.2011-1066

Crush J, Frayne B, McLachlan M (2011) Rapid urbanization and the nutrition transition in Southern Africa. Urb Food Secur Netw Series 7:1–49. Queen's University and African Food Security Urban Network (AFSUN), Kingston and Cape Town. Available https://fsnnetwork.org/sites/default/files/rapid_urbanization_and_the_nutrition.pdf. Accessed 7 April 2020

Curtis BM, O'Keefe JH Jr (2002) Understanding the Mediterranean diet: could this be the new "gold standard" for heart disease prevention? Postgrad Medic 112(2):35–45. https://doi.org/10.3810/pgm.2002.08.1281

D'Acampo G (2015) Gino's Islands in the Sun: 100 recipes from Sardinia and Sicily to enjoy at home. Hodder & Stoughton, London

Datar A, Nicosia N (2012) Junk food in schools and childhood obesity. J Pol Anal Manag 31(2):312–337. https://doi.org/10.1002/pam.21602

David E (2002) A book of Mediterranean food. New York Review of Books Inc., New York

Davis C, Bryan J, Hodgson J, Murphy K (2015) Definition of the Mediterranean diet; a literature review. Nutrients 7(11):9139–9153. https://doi.org/10.3390/nu7115459

De Batlle J, Garcia-Aymerich J, Barraza-Villarreal A, Antó JM, Romieu I (2008) Mediterranean diet is associated with reduced asthma and rhinitis in Mexican children. Allerg 63(10):1310–1316. https://doi.org/10.1111/j.1398-9995.2008.01722.x

De Graaf C (2005) Sensory responses, food intake and obesity. In: Mela DJ (ed) Food, diet and obesity. Woodhead Publishing Ltd., Cambridge, and CRC Press LLc, Boca Raton, pp137–159

De Souza GC, Dos Santos CTB, Andrade AA, Alves L (2015) Street food: analysis of hygienic and sanitary conditions of food handlers. Cienc Saude Colet 20(8):2329–2338

de Suremain CÉ (2016) The never-ending reinvention of 'traditional food'. In: Sébastia B (ed) Eating traditional food: politics, identity and practices. Routledge, Abingdon

Dedehayir O, Smidt M, Riverola C, Velasquez S (2017) Unlocking the market with vegan food innovations. In: Proceedings of the International Society for Professional Innovation Management (ISPIM) conference, Manchester, pp 1–13

Delgado AM, Almeida MDV, Parisi S (2017) Chemistry of the Mediterranean diet. Springer International Publishing, Cham. https://doi.org/10.1007/978-3-319-29370-7

Dittrich C (2017) Street food, food safety and sustainability in an emerging mega city: insights from an empirical study in Hyderabad, India. In: Xaxa V, Saha D, Singha R (eds) Work, institutions and sustainable livelihood. Palgrave Macmillan, Singapore

Dixon HG, Scully ML, Wakefield MA, White VM, Crawford DA (2007) The effects of television advertisements for junk food versus nutritious food on children's food attitudes and preferences. Soc Sci Med 65(7):1311–1323. https://doi.org/10.1016/j.socscimed.2007.05.011

Eames I, Khaw PT, Bouremel Y (2016) How to make the perfect pancake. Math Today 52(1):26–29

Erkaya T, Başlar M, Şengül M, Ertugay MF (2015) Effect of thermosonication on physicochemical, microbiological and sensorial characteristics of ayran during storage. Ultrason Sonochem 23:406–412. https://doi.org/10.1016/j.ultsonch.2014.08.009

FAO (1997) Street foods. FAO Food and Nutrition Paper N. 63. Report of an FAO Technical Meeting on Street Foods, Calcutta, India, 6–9 November 1995. Food and Agriculture Organization of the United Nations (FAO). Available https://www.fao.org/3/W4128T/W4128T00.htm/. Accessed 7 April 2020

FAO (2009) Good hygienic practices in the preparation and sale of street food in Africa. Food and Agriculture Organization of the United Nations (FAO), Rome. Available https://www.fao.org/3/a0740e/a0740e00.pdf. Accessed 9 April 2020

Fellows P, Hilmi M (2012) Selling street and snack foods. FAO Diversification booklet number 18. Rural Infrastructure and Agro-Industries Division, Food and Agriculture Organization of the United Nations (FAO), Rome. Available https://www.fao.org/docrep/015/i2474e/i2474e00.pdf. Accessed 7 April 2020

Gambella F, Paschino F, Cocciu P, Fadda M (2008) Analysis of work time and workers capacities in "culurgione's" production. In: Proceedings of the international conference "Innovation technology to empower safety, health and welfare in agriculture and agro-food systems", 15–17 September 2008, Ragusa, Italy

Giammanco GM, Pepe A, Aleo A, D'Agostino V, Milone S, Mammina C (2011) Microbiological quality of Pecorino Siciliano" primosale" cheese on retail sale in the street markets of Palermo, Italy. New Microbiol 34(2):179–185

Government of Canada (2020) Community-based measures to mitigate the spread of coronavirus disease (COVID-19) in Canada. www.canada.ca. Available https://www.canada.ca/en/public-health/services/diseases/2019-novel-coronavirus-infection/health-professionals/public-health-measures-mitigate-covid-19.html. Accessed 14 April 2020

Haddad MA, Abu-Romman S, Obeidat M, Iommi C, El-Qudah J, Al-Bakheti A, Awaisheh S, Jaradat DMM (2020a) Phenolics in Mediterranean and Middle East important fruits. J AOAC Int, in press

Haddad MA, Obeidat M, Al-Abbadi A, Shatnawi MA, Al-Shadaideh A, Al-Mazra'awi MI , Iommi C, Dmour H, Al-Khazaleh JM (2020b) Herbs and medicinal plants in Jordan. J AOAC Int, in press

Harris JB (2003) Beyond Gumbo: Creole fusion food from the Atlantic rim. Simon & Schuster, New York

Hawkes C, Harris J, Gillespie S (2017) Urbanization and the nutrition transition. Glob Food Pol Rep 4:34–41. https://doi.org/10.2499/9780896292529_04

Hayes A (2020) Coronavirus: food banks 'run low on basics' as donations fall and shoppers stock-pile. Sky UK. Available https://news.sky.com/story/coronavirus-food-banks-run-low-on-basics-as-donations-fall-and-shoppers-stockpile-11954815. Accessed 14 April 2020

Heinzelmann U (2014) Beyond bratwurst: a history of food in Germany. Reaktion Books Ltd., London

Helou A (2006) Mediterranean street food: stories, soups, snacks, sandwiches, barbecues, sweets, and more from Europe, North Africa, and the Middle East. Harper Collins Publishers, New York

Hoehl S, Rabenau H, Berger A, Kortenbusch M, Cinatl J, Bojkova D, Behrens P, Böddinghaus B, Götsch U, Naujoks F, Neumann P, Schork J, Tiarks-Jungk P, Walczok A, Eickmann M, Vehreschild MJGT, Kann G, Wolf T, Gottschalk R, Ciesek S (2020) Evidence of SARS-CoV-2 infection in returning travelers from Wuhan, China. New Engl J Med 382(13):1278–1280. https://doi.org/10.1056/nejmc2001899

Hopping BN, Erber E, Mead E, Sheehy T, Roache C, Sharma S (2010) Socioeconomic indicators and frequency of traditional food, junk food, and fruit and vegetable consumption amongst Inuit adults in the Canadian Arctic. J Human Nutr Diet 23:51–58. https://doi.org/10.1111/j.1365-277X.2010.01100.x

Hughner RS, McDonagh P, Prothero A, Shultz CJ, Stanton J (2007) Who are organic food consumers? A compilation and review of why people purchase organic food. J Cons Behav Int Res Rev 6(2–3):94–110. https://doi.org/10.1002/cb.210

INFOSAN (2010) Basic steps to improve safety of street-vended food. International Food Safety Authorities Network (INFOSAN) Information Note No. 3/2010—Safety of street-vended food, 30 June 2010. World Health Organization (WHO), Geneva. Available https://www.who.int/foodsafety/fs_management/No_03_StreetFood_Jun10_en.pdf. Accessed 15 April 2020

Keys A (1995) Mediterranean diet and public health: personal reflections. Am J Clin Nutr 61(6):1321S-1323S. https://doi.org/10.1093/ajcn/61.6.1321S

Khairuzzaman Md, Chowdhury FM, Zaman S, Mamun AA, Bari MdL (2014) Food safety challenges towards safe, healthy, and nutritious street foods in Bangladesh. Int J Food Sci Article ID 483519:1–9. https://doi.org/10.1155/2014/483519

Kollnig S (2020) The 'good people' of Cochabamba city: ethnicity and race in Bolivian middle-class food culture. Lat Am Caribb Ethn Stud 15(1):23–43. https://doi.org/10.1080/17442222.2020.1691795

Kowalczyk A, Korpysz A (2020) Restaurants and bars in the outer city. In: Kowalczyk A, Derek M (eds) Gastronomy and urban space. The Urban Book Series. Springer, Cham. https://doi.org/10.1007/978-3-030-34492-4_2

Kraig B, Sen CT (eds) (2013) Street food around the world: an encyclopedia of food and culture: an encyclopedia of food and culture. Abc-Clio, Santa Barbara

Lang T (1999) The complexities of globalization: the UK as a case study of tensions within the food system and the challenge to food policy. Agric Hum Values 16:169–185. https://doi.org/10.1023/A:1007542605470

Lappé FM, Collins J, Rosset P (1998) World hunger: twelve myths, 2nd edn. Grove Press, New York

Larcher C, Camerer S (2015) Street food. Temes de Dissen 31:70–83. Available https://core.ac.uk/download/pdf/39016176.pdf. Accessed 9 April 2020

Lee M, Choi Y, Quilliam ET, Cole RT (2009) Playing with food: content analysis of food advergames. J Consum Aff 43(1):129–154. https://doi.org/10.1111/j.1745-6606.2008.01130.x

Lee R (2017) Changing Hong Kong street food culture: an analysis of the Hong Kong food truck pilot scheme. Henry Luce China—Environment Program, Occidental College, 16 August 2017

Liakopoulos L, Furman G, Bonday LS, Hatzinikolaou D, Liu F (2015) Effect of a new high adhesion creping technology on machine runnability and tissue quality. Appita Technol Innov Manuf Environ 68(3):246–252

Liu Z, Zhang G, Zhang X (2014) Urban street foods in Shijiazhuang city, China: current status, safety practices and risk mitigating strategies. Food Control 41(1):212–218. https://doi.org/10.1016/j.foodcont.2014.01.027

Long-Solís J (2007) A survey of street foods in Mexico city. Food Foodways 15(3–4):213–236. https://doi.org/10.1080/07409710701620136

Lukić R (2011) Estimates of economic performance of organic food retail trade. Econ Res 24(3):157–169. https://doi.org/10.1080/1331677X.2011.11517474

Maiatsky M (2009) Pirojki à la bolognaise: Le miroir russe de l'université européenne. Multitudes 39(4):100–108. https://doi.org/10.3917/mult.039.0100

Manning JA (2009) Constantly containing. Dissertation, West Virginia University

Marangon F, Tempesta T, Troiano S, Vecchiato D (2016) Toward a better understanding of market potentials for vegan food. A choice experiment for the analysis of breadsticks preferences. Agric Agric Sci Proced 8:158–166. https://doi.org/10.1016/j.aaspro.2016.02.089

Maranzano B (2014) Lo sviluppo del fenomeno "street food": il cibo di strada a Palermo ieri e oggi. Dissertation, University of Pisa, Italy

Marcus E (2000) Vegan: the new ethics of eating. McBooks Press Inc., Ithaca

Martin V (2020) What impact could the coronavirus epidemic have on agriculture and food security? https://www.chinadaily.com.cn. Available https://www.chinadaily.com.cn/a/202002/24/WS5e53af6fa310128217279e6d.html. Accessed 14 April 2020

Matalas AL, Zampelas A, Stavrinos V, Wolinsky I (2001) The Mediterranean diet. Constituents and health promotion. CRC Press, Boca Raton, pp 46–66

Mchiza Z, Hill J, Steyn N (2014) Foods currently sold by street food vendors in the Western Cape, South Africa, do not foster good health. In: Sanford MG (ed) Fast foods: consumption patterns, role of globalization and health effects. Nova Science Publishers Inc., Hauppauge, pp 91–118

McHugh MR (2015) Modern Palermitan markets and street food in the Ancient Roman World. Conference paper, the Oxford Symposium on Food and Cookery, St. Catherine's College, Oxford University

Menezes F (2001) Food sovereignty: a vital requirement for food security in the context of globalization. Develop 44(4):29–33. https://doi.org/10.1057/palgrave.development.1110288

Messer E (1994) Food wars: hunger as a weapon of war in 1993. In: Uvin P (ed) The hunger report 1993. Yverdon, Switzerland, and Gordon and Breach Science Publishers, Philadelphia

Messer E (1996) Hunger and human rights, 1989–1994. In: Messer E, Uvin P (eds) The hunger report 1995. Yverdon, Switzerland, and Gordon and Breach Science Publishers, Philadelphia

Messer E, Cohen MJ (2001) Conflict and food insecurity. In: Diaz-Bonilla E, Robinson S (eds) Shaping globalization for poverty alleviation and food security. 2020 Vision Focus No. 8, Brief No. 12. International Food Policy Research Institute, Washington, D.C.

Messer E, Cohen MJ (2007) Conflict, food insecurity and globalization. Food Cult Soc 10(2):297–315. https://doi.org/10.2752/155280107X211458

Messer E, Uvin P (1996) Food wars: hunger as a weapon in 1994. In: Messer E, Uvin P (eds) The hunger report 1995. Yverdon, Switzerland, and Gordon and Breach Science Publishers, Philadelphia

Mlekuz J (2015) Burek: a culinary metaphor. Central European University Press, Budapest and New York

Mone I, Bulo A (2012) Total fats, saturated fatty acids, processed foods and acute coronary syndrome in transitional Albania. Mater Soc Med 24(2):91–93. https://doi.org/10.5455/msm.2012.24.91-93

Morton PE (2014) Tortillas: a cultural history. University of New Mexico Press, Albuquerque

Nasir VA, Karakaya F (2014) Consumer segments in organic foods market. J Cons Market 31(4):263–277. https://doi.org/10.1108/JCM-01-2014-0845

Nicolay N, McDermott R, Kelly M, Gorby M, Prendergast T, Tuite G, Coughlan G, McKeown P, Sayers G (2011) Potential role of asymptomatic kitchen food handlers during a food-borne outbreak of norovirus infection, Dublin, Ireland, March 2009. Eurosurveill 16(30):19931. https://doi.org/10.2807/ese.16.30.19931-en

Olsen WK, Warde A, Martens L (2000) Social differentiation and the market for eating out in the UK. Int J Hosp Manag 19(2):173–190. https://doi.org/10.1016/S0278-4319(00)00015-3

O'Mahony JM (2004) Designing molecularly imprinted polymers for the analysis of the components of complex matrices. Dissertation, Dublin City University

Omemu AM, Aderoju ST (2008) Food safety knowledge and practices of street vendors in the city of Abeokuta, Nigeria. Food Control 19(4):396–402. https://doi.org/10.1016/j.foodcont.2007.04.021

Oosterveer P, Sonnenfeld DA (2012) Food, globalization and sustainability. Earthscan, Abingdon and New York

Parente G (2007) Cibo veloce e cibo di strada. Le tradizioni artigianali del fast-food in Italia alla prova della globalizzazione. Storicamente 3, 2. https://doi.org/10.1473/stor389

Parisi S (2012) Food industry and food alterations. The user-oriented approach. Smithers Rapra Technologies, Shawsbury

Parisi S (2013) Food industry and packaging materials—performance-oriented guidelines for users. Smithers Rapra Technologies, Shawsbury

Parisi S (2019) Analysis of major phenolic compounds in foods and their health effects. J AOAC Int 102(5):1354–1355. https://doi.org/10.5740/jaoacint.19-0127

Parisi S (2020) Characterization of major phenolic compounds in selected foods by the technological and health promotion viewpoints. J AOAC Int, in press. https://doi.org/10.1093/jaoacint/qsaa011

Patel K, Guenther D, Wiebe K, Seburn RA (2014) Promoting food security and livelihoods for urban poor through the informal sector: a case study of street food vendors in Madurai, Tamil Nadu, India. Food Sec 6(6):861–878. https://doi.org/10.1007/s12571-014-0391-z

Petrenya N, Rylander C, Brustad M (2019) Dietary patterns of adults and their associations with Sami ethnicity, sociodemographic factors, and lifestyle factors in a rural multiethnic population of northern Norway-the SAMINOR 2 clinical survey. BMC Pub Health19(1):1632. https://doi.org/10.1186/s12889-019-7776-z

Pilcher JM (2017) Planet taco: a global history of Mexican food. Oxford University Press, Oxford

Pitte JR (1997) Nascita e diffusione dei ristoranti. In: Flandrin JL, Montanari M (eds) Storia dell'alimentazione, Laterza, Roma and Bari

Privitera D (2015) Street food as form of expression and socio-cultural differentiation. In: Proceedings of the 12th PASCAL international observatory conference, Catania

Puckette M, Hammack J (2015) Wine folly: the essential guide to wine. Avery, New York

Raak N, Dürrschmid K, Rohm H (2020) Textural characteristics of German foods: the German Würstchen. In: Nishihari K (ed) Textural characteristics of world foods. Wiley, Hoboken. https://doi.org/10.1002/9781119430902.ch23

Raynolds LT (2004) The globalization of organic agro-food networks. World Develop 32(5):725–743. https://doi.org/10.1016/j.worlddev.2003.11.008

Rohm H (1990) Consumer awareness of food texture in Austria. J Texture Stud 21(3):363–374. https://doi.org/10.1111/j.1745-4603.1990.tb00485.x

Romero-Gwynn E, Nicholson Y, Gwynn D, Raynard H, Kors N, Agron P, Fleming J, Sreenivasan L (1997) Refugees of former Soviet Union slowly adopt US diet. Calif Agric 51(6):24–28. https://doi.org/10.3733/ca.v051n06p24

Rosales S (2014) "This street is essentially Mexican": an oral history of the Mexican American community of Saginaw, Michigan, 1920–1980. Mich Hist Rev 40(2):33–62. https://doi.org/10.5342/michhistrevi.40.2.0033

Sabbithi A, Reddi SGDNL, Naveen Kumar R, Bhaskar V, Subba Rao GM, Rao VS (2017) Identifying critical risk practices among street food handlers. Brit Food J 119(2):390–400. https://doi.org/10.1108/BFJ-04-2016-0174

Sacks G, Veerman JL, Moodie M, Swinburn B (2011)'Traffic-light'nutrition labelling and 'junk-food'tax: a modelled comparison of cost-effectiveness for obesity prevention. Int J Obes 35(7):1001–1009. https://doi.org/10.1038/ijo.2010.228

Scholliers P, Van Molle L (eds) (2005) Land, shops and kitchens: technology and the food chain in twentieth-century Europe. Brepols Publishers, Turnhout. https://doi.org/10.1484/M.CORN-EB. 5.105946

Seyfang G (2009) The new economics of sustainable consumption. Seeds of change. Palgrave MacMillan, London

Sezgin AC, Sanlier N (2016) Street food consumption in terms of the food safety and health. J Hum Sci 13(3):4072–4083

Shadman Z, Poorsoltan N, Akhoundan M, Larijani B, Soleymanzadeh M, Zhand CA, Rohani ZAS, Nikoo MK (2014) Ramadan major dietary patterns. Iran Red Crescent Med J 16(9):e16801. https://doi.org/10.5812/ircmj.16801

Shaw N (2020) Food banks starting to run out because of coronavirus panic buying. Hull Daily Mail, UK. Available https://www.hulldailymail.co.uk/news/uk-world-news/food-banks-starting-run-out-3937595. Accessed 14 April 2020

Sheen B (2010) Foods of Egypt. Greenhaven Publishing LLC, New York

Silverman C (2012) Education, agency, and power among Macedonian Muslim Romani Women in New York City. Signs J Women Cult Soc 38(1):30–36. https://doi.org/10.1086/665803

Simopoulus AP, Bhat RV (2000) Street foods. Karger AG, Basel

Soldatou N, Nerantzaki A, Kontominas MG, Savvaidis IN (2009) Physicochemical and microbiological changes of "Souvlaki"—a Greek delicacy lamb meat product: evaluation of shelf-life using microbial, colour and lipid oxidation parameters. Food Chem 113(1):36–42. https://doi.org/10.1016/j.foodchem.2008.07.006

Sørensen S, Markedal KE, Sørensen JC (2018) Food, nutrition, and health in Denmark (including Greenland and Faroe Islands). Nutr Health Asp Food Nord Ctries 2018:99–125. https://doi.org/10.1016/B978-0-12-809416-7.00004-4

Souza LGS, Atkinson A, Montague B (2020) Perceptions about veganism. The Vegan Society, Birmingham

Sperling D (2010) Food law, ethics, and food safety regulation: roles, justifications, and expected limits. J Agric Environ Eth 23(3):267–278. https://doi.org/10.1007/s10806-009-9194-1

Stengs IL (2013) Dutch treats. Summer markets and the festive everydayness of Dutch fast food. Etnofoor, The Netherlands Now 25(2):145–158

Steven QA (2018) Fast food, street food: western fast food's influence on fast service food in China. Dissertation, Duke University, Durham

Steyn NP, Labadarios D (2011) Street foods and fast foods: how much do South Africans of different ethnic groups consume? Ethnic Dis 21(4):462–466

Steyn NP, Mchiza Z, Hill J, Davids YD, Venter I, Hinrichsen E, Opperman M, Rumbelow J, Jacobs P (2013) Nutritional contribution of street foods to the diet of people in developing countries: a systematic review. Pub Health Nutr 17(6):1363–1374. https://doi.org/10.1017/S1368980013001158

Sujatha T, Shatrugna V, Rao GN, Reddy GCK, Padmavathi KS, Vidyasagar P (1997) Street food: an important source of energy for the urban worker. Food Nutr Bull 18(4):1–5. https://doi.org/10.1177/156482659701800401

Swinnen JF, Maertens M (2007) Globalization, privatization, and vertical coordination in food value chains in developing and transition countries. Agric Econ 37:89–102. https://doi.org/10.1111/j.1574-0862.2007.00237.x

Thakur CP, Mehra R, Narula C, Mahapatra S, Tj K (2013) Food safety and hygiene practices among street food vendors in Delhi, India. Int J Curr Res 5(11):3531–3534

Tinker I (1999) Street foods into the 21st Century. Agric Human Val 16:327–333

Uğurlu K (2015) The impacts of Balkan cuisine on the gastronomy of thrace region of Turkey. In: Csobán K, Könyves E (eds) Gastronomy and culture. University of Debrecen, Debrecen, pp 43–66

Vanschaik B, Tuttle JL (2014) Mobile food trucks: California EHS-Net study on risk factors and inspection challenges. J Environ Health 76(8):36–37. Gale Academic OneFile. Available https://go.gale.com/ps/anonymous?id=GALE%7CA365689844&sid=googleScholar&v=2.1&it=r&linkaccess=abs&issn=00220892&p=AONE&sw=w. Accessed 7 April 2020.

Vareltzis P, Adamopoulos K, Stavrakakis E, Stefanakis A, Goula AM (2016) Approaches to minimise yoghurt syneresis in simulated tzatziki sauce preparation. Int J Dairy Technol 69(2):191–199. https://doi.org/10.1111/1471-0307.12238

Verbeke WA, Viaene J (2000) Ethical challenges for livestock production: Meeting consumer concerns about meat safety and animalwelfare. J Agric Environ Eth 12(2):141–151. https://doi.org/10.1023/A:1009538613588

Webb RE, Hyatt SA (1988) Haitian street foods and their nutritional contribution to dietary intake. Ecol Food Nutr 21(3):199–209. https://doi.org/10.1080/03670244.1988.9991033

Weckroth K (2018) Towards plant-based food consumption practices: activity focus group study. Dissertation, University of Tampere. Available https://trepo.tuni.fi/handle/10024/103610. Accessed 14 April 2020

WHO (2020) Emergencies—disease maps of countries affected by food insecurity and famine. World Health Organization (WHO), Geneva. Available https://www.who.int/emergencies/famine/disease-maps/en/. Accessed 14 April 2020

WHO Department of Food Safety, Zoonoses and Foodborne Diseases (2006) Five keys to safer food manual. World Health Organization (WHO), Geneva. Available https://apps.who.int/iris/bitstream/10665/43546/1/9789241594639_eng.pdf?ua=1. Accessed 15 April 2020

Wiles NJ, Northstone K, Emmett P, Lewis G (2009) 'Junk food' diet and childhood behavioural problems: results from the ALSPAC cohort. Eur J Clin Nutr 63(4):491–498. https://doi.org/10.1038/sj.ejcn.1602967

Wilkins J, Hill S (2009) Food in the ancient world. Blackwell Publishing Ltd, Maiden, Oxford, and Carlton

Wilkinson J (2004) The food processing industry, globalization and developing countries. Electron J Agric Develop Econ 1(2):184–201

Willett WC, Sacks F, Trichopoulou A, Drescher G, Ferro-Luzzi A, Helsing E, Trichopoulos D (1995) Mediterranean diet pyramid: a cultural model for healthy eating. Am J Clin Nutr 61(6):1402S-1406S. https://doi.org/10.1093/ajcn/61.6.1402S

Yaniv G, Rosin O, Tobol Y (2009) Junk-food, home cooking, physical activity and obesity: the effect of the fat tax and the thin subsidy. J Pub Econ 93(5–6):823–830. https://doi.org/10.1016/j.jpubeco.2009.02.004

Yiridoe EK, Bonti-Ankomah S, Martin RC (2005) Comparison of consumer perceptions and preference toward organic versus conventionally produced foods: a review and update of the literature. Renew Agric Food Sys 20(4):193–205. https://doi.org/10.1079/RAF2005113

Zimmerman FJ, Bell JF (2010) Associations of television content type and obesity in children. Am J Pub Health 100(2):334–340. https://doi.org/10.2105/AJPH.2008.155119

Chapter 2
Palermo's Street Foods. The Authentic *Arancina*

Abstract Street food is synonym of historical dominations and business methods at least: this discussion should concern non-technical disciplines such as history, architectural design, marketing, and other ambits with relation to chemistry, microbiology, hygiene, and technological features of foods and beverages. More than two billion consumers worldwide prefer street foods. It may be assumed in certain ambits and social areas that the consumption of street foods is associated with a social elevation in terms of cultural defense of traditions, the spreading of different versions worldwide, and the consequent cultural 'contamination', with some exceptions. Sicily is a geographically limited area in the Mediterranean Basin. Its history has a long number of cultural contaminations, and several Sicilian 'street foods' appear to be a localised characteristic, so that the typical Palermo inhabitants consider these specialities as the pride of the area and the demonstration of cultural identity. For this reason, the book has been also named 'the Palermo case study'. The present Chapter concerns one of these specialities, the '*arancina*', and the alternative Sicilian version—the Catania's or Western side '*arancino*'—by different viewpoints including history, possible 'authenticity' features, identification of raw materials, preparation procedures, concomitant alternative recipes, and nutrition facts.

Keywords Arabic cuisine · *Arancina* · *Arancino* · Frying · Rice · Sicily · Street food

Abbreviations

CNC	Consiglio Nazionale dei Chimici
CMYK	Cyan, Magenta, Yellow, Black
EVOO	Extra-virgin olive oil
MiPAAF	Ministero delle politiche agricole alimentari e forestali
SF	Street food
TSG	Traditional speciality guaranteed
UNED	Universidad Nacional de Educación a Distancia

2.1   Sicilian Street Food and Localised Traditions

Street food (SF) is a synonym of historical dominations, business methods (selling strategies 'on-the-road'), and more in general of anthropic activities. In other words, it could be assumed that the simple description of SF specialities should concern non-technical disciplines such as history, architectural design, marketing, and other ambits without relation to chemistry at least.

Actually, this hypothesis is surely incorrect. Many scientific papers are based on the obvious correlation between SF and human safety, and this argument has to be further correlated with chemical, microbiological, and technological features of foods and beverages. The importance of humanistic disciplines has certainly an impact when speaking of folk traditions. The 'success' and the temporal persistence of certain traditions has to be correlated with current conditions of the food and beverage market on the one side, and original reasons which have caused the production and the spreading of similar dishes and habits around the world.

As mentioned in Chap. 1, SF specialities have some key points if compared with similar 'fast-food' offers, at least in certain cultural areas such as Italy, and especially Sicily. Economic convenience is certainly the best reason (Chukuezi 2010; Fellows and Hilmi 2012): one single piece of soft or salted SF is reported to be purchased in Palermo with very small prices (Parente 2007). For this and other reasons, more than two billions of consumers worldwide prefer SF: these offers are promising choices if compared with other foods.

From the nutritional angle, SF may be good enough when speaking of energy intakes and bioavailable protein contents (Alfiero et al. 2017; Allison 2018; Dittrich 2017; Lee 2017; Rosales 2014). Interestingly, the possible social discrimination which could be perceived by SF consumers is not always demonstrable. On the contrary, it may be assumed in certain ambits and social areas that the consumption of SF is associated with a social elevation in terms of cultural defence (Di Giorgi 2017; Parente 2007).

Another important fact should be considered with attention because of the difficult repetition of the same phenomenon worldwide. The fragmentation of the current food and beverage market has been reported many times in the scientific literature, with relation to certain products and related features (Alfiero et al. 2017; Alves da Silva et al. 2014; FAO 2009; Hopping et al. 2010; Long-Solís 2007; Maranzano 2014; Privitera 2015; Simopoulus and Bhat 2000; Steyn and Labadarios 2011; Steyn et al. 2013; Tinker 1999). It has to be noted also that many regional or national dishes have spread across the world for different reasons with partially predictable results such as mixtures between Italian *pizza* and *kebab* at the same time. Other reported association—the Russian-style *pirozhki* dish with *bolognese* (Italian-style) sauce (Maiatsky 2009) or the Balkan-style food named '*pizza/burek*' (Brown 2018; Kraig and Sen 2013; Silverman 2012)—may be mentioned at the same time. Actually, the phenomenon of cultural 'contamination' is commercially surprising because of the possible offer of 'traditional' recipes with one or more ingredients showing a well-known commercial brand (Toktassynova and Akbaba 2017). Anyway, these reports should demonstrate that SF can be an expression of contamination… with some exceptions.

Sicily is a geographically limited area in the Mediterranean Basin. Its history has a long number of cultural contaminations: Phoenician, Roman, Byzantine, Muslim, Norman, Suevian, Angevin, Aragonese, Spanish, and the Bourbon periods have to be mentioned. Each of these dominations and civilisations has given something to Sicily, including the persistence of certain surnames along centuries, the reminiscence of several historical facts with huge importance at the European level, and also several traces in culinary traditions. Perhaps, it might be assumed that there are many civilisations in Sicilian history… and consequently, Sicilians could be no longer available to accept further contaminations at least in their culinary traditions. It is only an interpretation, but this reflection could also explain why certain Sicilian dishes are not easily contaminated with other ingredients or materials, similarly to above-mentioned examples. Many food specialities in Sicily should be discussed in this and in other ambits, from hard-boiled octopus, *cicireddu* (fried little fish), and fried *supplì* (rice croquettes filled with tomatoes, mozzarella cheese, and meat) (Parente 2007). However, the argument is extremely broad, and it should be evaluated critically in more than one single book.

The prominence of fried SF in Sicily, and in the Palermo area in particular, should be considered as a peculiar heritage of ancient civilisation, above all the Arab tradition (Di Giorgi 2017; Parente 2007). The typical Palermo inhabitants consider these specialities as the pride of the area and the demonstration of cultural identity. What about these peculiarities? Probably, the reasons have to be researched in ancient history, but the nature of these foods may be also helpful. In fact, each food or beverage has its reasons inherent in its composition and production, such as a general cheese has its main meaning as 'preserved milk' (if compared with perishable milk). The description of certain SF could be very useful with the aim of finding historical and marketing reasons for the success of SF in Sicily and abroad. For this reason, the book has been also named 'the Palermo case study'.

The present Chapter concerns one of the most known SF specialities in Sicily, and in the Palermo area: the '*arancina*'. Actually, it should be admitted that the more the Sicilian cities with a notable cuisine tradition, the more the possible versions of a typical recipe … and this situation is peculiar when speaking of *arancina* versions. In fact, by the historical viewpoint, there are two possible and different types of the same food at least, although these typologies are geographically close (only 166 kms in the same Region). The differences concern not only the recipe and preparation/cooking procedures, but also the same shape and the name (Figs. 2.1 and 2.2), with interesting and sometimes inexplicable rivalries (Bartolotta 2010; Camilleri 2018; Iannizzotto 2016; Marino et al. 2010; Ricci 2014).

Fig. 2.1 By the historical viewpoint, there are two possible and different types of 'rice balls' in Sicily, although these typologies are geographically close (only 166 km in the same Region…). The differences concern not only the recipe and preparation/cooking procedures, but also the same shape and the name, with interesting and sometimes inexplicable rivalries (Camilleri 2018; Di Giorgi 2017; Marino et al. 2010). The Palermo's version or Western type is named '*arancina*'

2.2   Sicilian Street Foods. One or More Specialities with Similar Names

The SF speciality commonly named '*arancina*' is found typically in the Sicilian capital city, Palermo, on the western side (Fig. 2.1). On the eastern side, there are different typologies of the 'same' product, typically produced and served in Messina, Ragusa, Siracusa provinces, and above all in the central province surrounding the Mount Etna: Catania. For this reason, the antagonist type is commonly intended as the Catania-style SF '*arancino*' (Fig. 2.2) while differences do not concern only the final vowel of the name (*o* against *a*…).

Anyway, and in spite of differences between the two typologies, it can be assumed that this SF category may be defined as a rice ball where Arabic cuisine has its weight and historical importance (Nessi 2018).

Basically, this SF is a rice ball filled with different ingredients, and subsequently breaded and fried (Marino et al. 2010). In detail (and with considerable difficulties because of the coexistence of two different and similar products in Palermo and in

Fig. 2.2 By the historical viewpoint, there are two possible and different types of 'rice balls' in Sicily, although these typologies are geographically close (only 166 km in the same Region…). The differences concern not only the recipe and preparation/cooking procedures, but also the same shape and the name, with interesting and sometimes inexplicable rivalries (Camilleri 2018; Di Giorgi 2017; Marino et al. 2010). The Catania's version or Eastern type (diffused widely in Sicilian eastern provinces) is named '*arancino*'

Catania), this food may be described with the following attributions (Marino et al. 2010):

(1) A fried rice ball,

(2) A diameter between 8 and 10 cm,

(3) Fillings: Italian-style *ragù* with chopped meat (Caldara 2013; da Vico et al. 2010; Guzzo 2014; Gwinner 2018), butter, mozzarella or other cheeses (Alves et al. 1996; Haddad and Parisi 2020), and normal or frozen green peas (Giannakourou and Taoukis 2003; Hansen et al. 2000; Hung and Thompson 1989; Mohan et al. 2014; Sorensen et al. 2003),

(4) The related name (*arancina* in Palermo, *arancino* in Catania) is apparently derived from the shape and the colour, both similar to an orange fruit (Cumbo 2000; Summerfield 2009; Tomaiuolo 2009; Wright 1996).

With reference to above-mentioned descriptions, a premise should be discussed when speaking of fillings, basic coating preparations with breadcrumbs, the shape, dimensions, and finally the name. Before starting with this regional discussion, it should be mentioned that a smaller version of *arancino* or *arancina* is often found as

a typical SF speciality outside Sicily. Street food vendors in Naples are accustomed
to prepare and serve fried, round, and small (if compared with Sicilian products)
rice balls, and the name is curiously the dialectal translation of 'rice balls': '*pall'e
riso*' (Marino et al. 2010). Anyway, this book is dedicated to the Sicilian version;
consequently, the description of the Neapolitan SF rice ball is not described and
discussed here.

2.2.1   *Arancina Versus Arancino. Shapes and Dimensions*

An important and certainly visible difference concerning Palermo and Catania (or
Western and Eastern varieties) is the shape. The *arancina* type (gender name: female,
in Italian language) corresponds to a spherical shape as above mentioned (Figs. 2.1
and 2.3), while *arancino* type (gender name: male, in Italian language) appears like
a cone with the vertex facing upwards (Figs. 2.2 and 2.4) (Maranzano 2014; Marino
et al. 2010).

Fig. 2.3 An important and certainly visible difference concerning *arancina* and *arancino* SF is
the shape. The *arancina* type (gender name: female, in Italian language) corresponds to a spherical
shape as above mentioned). This picture shows the 'red' (with meat) and 'white' (with butter and
ham) versions (Maranzano 2014; Marino et al. 2010)

Fig. 2.4 An important and certainly visible difference concerning *arancina* and *arancino* SF is the shape. The *arancino* type (gender name: male, in Italian language) appears a cone with the vertex facing upwards (Maranzano 2014; Marino et al. 2010) similarly to the Mount Etna (Catania province)

These differences appear to be based on historical reasons. It has been reported that Palermo's spherical version depends on the ancient influence of Arabic cuisine (Nessi 2018). In those years (827–1091 A.D.), the Arabs favoured different fruits, including the bitter orange (*Citrus aurantium*) and rice (Specie: *Oryza sativa*) (Pizzuto Antinoro 2002; Pensovecchio 2017). According to several researchers, the circumstance that Palermo was the capital city of the ancient Sicily Emirate (Johns 2002) could have favoured not only the birth of an orange-like food in the whole island, but also the name similar to the Italian meaning and gender (D'Ignoti 2019). In addition, the Arab name of oranges is *naranji* (from the Persian   , *nārang*, while the corresponding and derived Spanish term is *naranja*). These factors may have determined the diffusion of the *arancina* name in the western areas of Sicily, and particularly in the Capital.

On the other side of Sicily, the popular belief that *arancino* represents Mount Etna has been repeatedly considered (Chef Rubio 2014). In general:

(1) The conic shape of this SF should be similar to a vulcan (Figs. 2.2 and 2.4).
(2) The exterior crisp layers should be similar to the upper surface of Mount Etna.
(3) The inner parts of *arancino*, containing chopped meat and red sauces such as *ragù*, should indicate the reddish lava exiting from the inner layers of vulcans.

(4) And, last but not least, the opening of hot *arancino* pieces should remember the inner and hot vapours exiting from the Mount Etna.

As above mentioned, the discussion concerning *arancina* and *arancino* types appears essentially historical. There are not irrefutable justifications for one or another reason.

In general, it may be assumed that the nature of SF can be ascribed to *arancina* and *arancino* because Arabs were accustomed to eat rice and seasoned saffron with meat and herbs, and the subsequent Sicilian domination may have turned this habit into a new 'portable' (take-away) and easily preservable food. It has been reported that crispy breading may have a similar preservative function, if compared with cooked rice without external coatings (Droga and Lo Bue 2009; Marino et al. 2010).

However, this discussion has been exhibited here because of the need for a differentiation between the two typologies, even if both versions are similar and essentially defined SF. This book concerns only Palermo's SF; consequently, the *arancino* version is not discussed in detail, except for differences concerning the visible and reliable composition of these foods.

2.2.2 *Arancina Versus Arancino. Preparation and Differences*

Certain differences between these rice balls exist also in relation to preparation and cooking procedures.

2.2.2.1 *Arancina* **Preparation (Western or Palermo's Version)**

Basically, the preparation of *arancina* types concerns the following steps (Fig. 2.5):

(a) Rice (*Oryza sativa* sub. *japonica*) (CNC 2014a, b) with round grains has to be cooked '*al dente*' (Italian expression meaning that the rice has to receive a compact enough texture and characteristic chewingness) (Crowhurst and Creed 2001; Sozer 2009; Suwannaporn and Linnemann 2008).

(b) Mixture step. Cooked *al dente* rice grains have to be mixed with animal fat (butter) and cheese (*caciocavallo*, mozzarella, *pecorino*, etc.).

(c) Temporary cooling. The obtained mixture has to remain on a cool surface (traditionally, marble surfaces).

(d) Disc formation. The operator uses the intermediate mixture for realising several discs, where the centre has to contain the final filling (Sect. 2.2.3).

(e) Ball formation and coating step. Discs have to be manually filled and closed with the aim of obtaining the spherical shape; subsequently, each ball is coated by simple contact with beaten egg mixtures and breadcrumbs.

(f) Browning step. The obtained and coated balls have to be fried in hot peanut oil (Sect. 2.3.4.3).

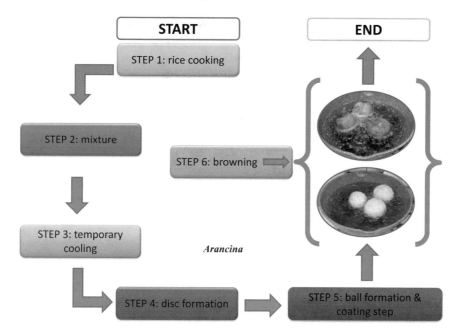

Fig. 2.5 Basically, the preparation of (red) *arancina* types concerns seven steps

In addition, it has to be considered that saffron is used when speaking of mixture step with the aim of obtaining a yellowish colour. Moreover, the use of saffron would allow a good and visible separation between the coating layers and inner rice mixture.

2.2.2.2 *Arancino* Preparation (Eastern or Catania's Version)

Basically, the preparation of *arancino* types concerns the following steps:

(a) Rice with round grains has to be cooked '*al dente*' (Italian expression meaning that the rice has to receive a compact enough texture and characteristic chewingness) (Crowhurst and Creed 2001; Sozer 2009).
(b) Mixture step. Cooked *al dente* rice grains have to be mixed with animal fat (butter) and cheese (*caciocavallo*, mozzarella, *pecorino*, etc.).
(c) Temporary cooling. The obtained mixture has to remain on a cool surface (traditionally, marble surfaces).
(d) Disc formation. The operator uses the intermediate mixture for realising several discs, where the centre has to contain the final filling (Sect. 2.2.3).
(e) Ball formation and coating step. Discs have to be manually filled and closed with the aim of obtaining the cone-like typical shape; subsequently, each ball is coated by simple contact with beaten egg mixtures and breadcrumbs.
(f) Frying step. The obtained and coated balls have to be fried in hot oil.

In addition, it has to be considered that saffron is not used in the mixture step, differently from Palermo's version. On the contrary, a prepared sauce is used instead of the more expensive saffron. In this version, the inner colour is reddish and there are no visible differences between the coating layers and inner rice mixture.

2.2.3 Arancina Versus Arancino. Fillings

An important step in the comprehension of these versions concerns fillings. Once more, it may be difficult to understand the reason(s) for some different filling compositions with less than 200 km of distance between the main centres of the Sicilian Island.

With reference to *arancina* (Palermo, western type), the most known and used fillings are (Marino et al. 2010).

(a) Red version: a mixture of *ragù* (meat sauce), green peas, and carrots; or
(b) White version: a mixture containing butter, mozzarella cheese (although other cheeses such as *pecorino* or *caciocavallo* types are described), and ham; or
(c) Mixture with spinach and mozzarella cheese.

On the eastern side of the Island, the traditional Catania's *arancino* contains also aubergines (one version) and Pistachio of Bronte (Marino et al. 2010).

It should be also noted that in recent years these SF have been subjected to several contaminations in terms of 'foreign' fillings. It has been mentioned that these SF are not easily found in association with other regional specialities, especially if non-Italian types. However, the insertion of salmon, swordfish, sausages, etc. corresponds always to a commercial mutation of original formulations, and it should be reported. In addition, these foods are not sweet products, while SF vendors may nowadays serve also new rice balls where cocoa replaces the original filling and sugar can replace breadcrumbs (Marino et al. 2010).

2.3 The Palermo Case Study and *Arancina* Foods. A Detailed Description

The production of *arancina* or *arancino* foods depends on the peculiar location (western or eastern sides of the Sicilian island), but also on the shared and agreed method(s). The problem is that generally there are many different interpretations of the same food type. As a result, the higher the number of different versions, the higher the complexity of a reliable description… It has to be also considered that this food has been officially recognised as one of the Italian traditional food products or 'prodotti agroalimentari tradizionali italiani' (PAT) of the Italian Ministry of Agricultural, Food and Forestry Policies (MiPAAF) (Ministero delle politiche agricole alimentari e forestali 2014). The confusion between *arancina* and *arancino* is evident here because this product is apparently described as the western *arancina* and with the name 'rice *arancino*' (the eastern version)…

This Chapter would give a simplified and reliable description of similar foods, and also mentioned with relation to other Palermo's SF (Chaps. 3, 4, and 5). With reference to the Palermo's *arancina*, the objective should be the description of an 'authentic way' to this product on the basis of selected references in the following way:

(a) Product typology,
(b) External appearance,
(c) Weight,
(d) Method of preparation,
(e) Raw materials and filling,
(f) Qualitative and approximate chemical composition of *arancina* SF.

Before proceeding in this way, an initial premise has to be considered. All the discussed data and results come from selected references with the aim of showing different visions of the same product. This reflection has to be taken into mind.

In 2014, a project concerning the application for registration of a 'traditional speciality guaranteed' (TSG), according to the Title III of the Regulation (EU) No. 1151/2012 of the European Parliament and of the Council of 21 November 2012 on quality schemes for agricultural products and foodstuffs has been proposed in Italy with relation to *arancina* foods (CNC 2014a, b). This document may be helpful when speaking of a general way to define the authentic *arancina* SF, with some additional discussions.

Before going on with the discussion, it should be mentioned that (Maranzano 2014) this proposal followed the creation of a '*City Brand Panormvs*' for the protection of Palermo's products. The TSG proposal aims at the creation of the '*Panormvs— street food*' brand specifically linked to the TGS product(s) and the realisation of an international SF festival in Palermo (Andolina 2019). At present, there is a 'Panormus Street Food' SF mobile caterer company currently working in the United Kingdom and serving also *arancina* SF. On the other side, there is not available information concerning the *arancina* TSG: there are not Italian applications in this ambit as TSG on 24 March 2020 (European Commission 2019, 2020).

2.3.1 Product Typology

According to the above-mentioned TSG proposal, the *arancina* SF (with this specific name, excluding the eastern *arancino* version) should be classified as follows:

– Class of product (according to Annex I f the above-mentioned Reg. 1152/2012, II.3): bread, pastry, cakes, confectionery, biscuits, and other baker's wares.

2.3.2   External Appearance and Shape

According to the TSG proposal, the *arancina* (with this specific name, excluding the eastern *arancino* version) would have the following basic features:

(a) Shape of the product. The food should be spherical as reported in the previous sections, with a diameter between 8 and 10 cm, and a weight between 150 and 200 g.

(b) The external appearance of the SF resembles a normal orange fruit, according to historical traditions and above-mentioned interpretations, with a visible orange-like colour.

(c) The product is generally described as a food obtained by means of the slow frying (browning or sautéing method) in hot oil of the spherical envelope (containing a heterogeneous filling). The browning method and the preparation of the whole food should assure the physical resistance (no ruptures) during the browning itself and after this step. Also, the oil used for this is defined as 'extra-virgin olive oil' (EVOO) when speaking of filling preparation (Sect. 2.3.4.2), and peanut oil with relation to the external envelope (Sect. 2.3.4.3). We will consider the term 'browning' in this book when speaking of slow and controlled frying in hot oil (because a certain colourimetric modification is needed). Alternatively, this frying technique can be named 'sautéing'. However, we will use this term conventionally even if the desired result does not concern only colours (such as in Chap. 1).

2.3.3   Weight

As above mentioned, the *arancina* SF should be spherical with a diameter between 8 and 10 cm, and a weight between 150 and 200 g. This range is probably determined by traditions and the need for obtaining a resistant structure (no ruptures during and after browning). Basile reported that 200 g should be the best result in this ambit (Basile 2015), but modern SF vendors are used to produce also oversized products tending to approximate weights of 500 g (Parente 2007). This fact has to be considered at present, especially when speaking of caloric intake and approximate composition.

2.3.4   Method of Preparation

The shown method of preparation is discussed here with relation o the 'red' version of *arancina* SF. It is mainly described on the basis of the above-mentioned TSG proposal, although other traditional references have been considered.

2.3.4.1 Rice Cooking

Rice is cooked in salted water (time: 15 ±1 min) until it can reach the *al dente* texture. Saffron has to be added with the aim of reaching a 'Cyan, Magenta, Yellow, Black' (CMYK) value such as 0-4-44-0 (CNC 2014a, b).

2.3.4.2 Filling and Shaping

The filling has to be prepared with raw materials as defined in Sect. 2.3.5. The detailed procedure has to be carried out for 40 min and the final result should have a 12–14°Brix desired result as dry matter. The mixture has to be placed until the mass will be cold enough to allow the subsequent shaping in a spherical way with prepared rice.

2.3.4.3 Final Frying

The intermediate *arancina* has to be fried in peanut oil with wheat flour and water (ratio: 1:1), temperature 170–180 °C, time: four minutes. The fried product ahs to be finally coated with breadcrumbs by simple contact (manually) and the aim of this procedure also concerns the elimination of excess oil from surfaces (CNC 2014a, b).

The browning procedure should be considered in relation to the type of used oil and a minimum recommended temperature. However, the colourimetric yield seems more important. It has been reported, with relation to the *arancino* version, that oil should reach 190 °C (Marino et al. 2010). As above mentioned, it appears that the optimal cooking/browning should be considered with relation to colourimetric yields only (the characteristic orange-like colour).

2.3.5 Raw Materials for Filling

This procedure is shown for 'red version' only and mainly based on the above-mentioned TSG proposal. All prepared ingredients (cow meat, onion, green peas, celery spines, diced carrots, and reconstituted tomato concentrate in water (8–10°Brix for dry matter) are mixed together after onion browning in EVOO, addition of peas, celery spines, and diced carrots. Meat has to be added and browning process has to be continued until the mixture has a light golden colour, and finally the tomato-reconstituted sauce has to be added. The browning process has to be continued for 40 min.

The filling is described as composed of chopped cow meat with green peas (minimum diameter: 5 mm), diced carrots, finely minced onions, celery spines, and reconstituted tomato concentrate (from double tomato concentrate). In this ambit, it has to be considered that *ragù* corresponds to mixed sauce and minced meat.

Consequently, ingredients can be defined as expressed in the proposal (six raw materials, excluding rice and breadcrumbs) or as *ragù*, peas, carrots, onions, and celery (four ingredients). In addition, the Sicilian preparation of homemade sauces contains onions. As a result, the number of composed ingredients could be reduced to three points.

With relation to allowed raw materials and ingredients, the following list should be considered, taking into account the difference between 'red' (meat and sauce-containing) and 'white' (butter and ham-containing) *arancina* (Cabibbo 2018; Maranzano 2014; Mele 2016; Parente 2007):

(a) Rice (also parboiled rice, on condition that contained starch sauce (the mass of starch molecules actively linking together rice grains) is not sensorially perceived,

(b) Cow meat (pork meat is also allowed if mixed with cow meat on condition that the fat/meat ratio allows the result of a soft mixture between the two meat types),

(c) Double tomato concentrate (to be reconstituted),

(d) Green peas,

(e) Carrots,

(f) Onions,

(g) Celery spines,

(h) EVOO,

(i) Peanut (groundnut) oil,

(j) Saffron,

(k) Breadcrumbs,

(l) Water,

(m) W*heat flour 00 (double zero)* (Balsari et al. 2013),

(n) Eggs.

The above-mentioned list may be challenging enough. For this reason, Table 2.1 shows the same ingredients in relation to the peculiar type of *arancina* SF. It has to be considered that suggested amounts are only one of the many possibilities when speaking of *arancina* SF. Only ingredients mentioned both in the TSG proposal and in other references have been considered, while possible ingredients such as black pepper are not placed here because of their existence in some of cited references only. It has to be considered that suggested amounts are only one of the many possibilities when speaking of *arancina* SF. These data are based on the interpolation of data from selected references (Cabibbo 2018; Maranzano 2014) only for a qualitative and semi-quantitative analysis. None of these advices may constitute a premise or guarantee for *arancina* authenticity: in other words, there are not reliable amounts to be shown and discussed/declared concerning 'authentic' *arancina* SF. The mentioned amounts do not correspond to the ingredient percentage in the final product.

With reference to filling, it should be noted that the butter version (with a pronounced oval shape) is not mentioned in the TSG proposal. Consequently, the TSG proposal would not apparently consider the butter version as 'authentic'. On the other hand, this SF is widely found in Palermo (Basile 2015). At the same time,

Table 2.1 Allowed ingredients for *arancina* SF

Ingredients	Meat sauce (red)*arancina*	Butter and ham (white) *arancina*	Notes (R is for red, W is for white)
Rice	YES	YES	Parboiled rice is also allowed on condition that contained starch sauce (the mass of starch molecules actively linking together rice grains) is not sensorially perceived Suggested amounts: 31.3% (R), 28.1% (W)
Cow meat	YES	NO	Suggested amount: 21.9% or 10.9% if in association with pork meat
Pork meat	YES	NO	Pork meat is allowed if mixed with cow meat on condition that the fat/meat ratio allows the result of a soft mixture between the two meat types) Suggested amount: 10.9%
Bechamel (white sauce, cream sauce)	NO	YES	Suggested amount: 14.0% Obtained from: butter, wheat flour, and milk (approximate composition: 8:1:1 w/w/w)
Mozzarella or *provola* cheese	YES	YES	Suggested amount: 3.8% (R), 6.7% grams (W)
Tomato concentrate	YES	NO	To be reconstituted in water (final dry matter concentration: 8–10°Brix) Suggested amount (as a reconstituted sauce): 15.6%
Green peas	YES	NO	Suggested amount: 3.1%
Carrots	YES	NO	Suggested amount: 2.7%
Onions	YES	NO	Suggested amount: 1.4%
Celery spines	YES	NO	–
EVOO	YES	YES	Suggested amount: *quantum satis*
Peanut (groundnut) oil	YES	YES	Suggested amount: *quantum satis*

(continued)

Table 2.1 (continued)

Ingredients	Meat sauce (red)*arancina*	Butter and ham (white) *arancina*	Notes (R is for red, W is for white)
Butter	NO	YES	Suggested amount: 3.4%
Ham	NO	YES	Suggested amount: 11.2%
Saffron	YES	YES	<0.01%
Breadcrumbs	YES	YES	Suggested amount: 6.3% (R), 11.2% (W)
Water	YES	YES	Suggested amount: 4.7% (R), 8.4% (W)
Salt	YES	YES	Suggested amount: *quantum satis*
W*heat flour 00 (double zero)* (Balsari et al. 2013)	YES	YES	Suggested amount: 4.7% (R), 8.4% (W)
Eggs	YES	YES	Suggested amount: 4.7% (R), 8.4% (W)

Two lists are considered with reference to meat and butter and ham versions (CNC 2014a, b; Maranzano 2014; Parente 2007). Only ingredients mentioned both in the TSG proposal and in other references have been considered, while possible ingredients such as black pepper are not placed here because of their existence in some of cited references only. Suggested amounts for preparation are only one of many possibilities when speaking of *arancina* SF. These data are only based on the interpolation of data from selected references (Anonymous 2020a; CNC 2014a, b; Di Giorgi 2017; Maranzano 2014; Parente 2007), for a qualitative and semi-quantitative analysis only. None of these advices may constitute a premise or guarantee for *arancina* authenticity: in other words, there are not reliable amounts to be shown and discussed/declared concerning 'authentic' *arancina* SF. The mentioned amounts do not correspond to the ingredient percentage in the final product

it should be considered the role of tomato sauce because of its introduction in Sicily after the Arab period. Substantially, and similarly to other traditions in the Island, the use of tomato sauces has probably modified the original recipe (Gentilcore 2010). This situation has to be considered because the two types are sometimes named 'red *arancina*' (meat and sauce) and 'white *arancina*' (butter, cheese, and ham).

Another difference should concern the use of cheeses. Some authors mention the use of *provola* cheese in the meat (*ragù*) version, while the TSG proposal does not appear to consider it. At the same time, the use of mozzarella cheese is reported in 'white' butter version (Parente 2007), while the possibility of *ragusano* cheese is admitted.

2.3.6 Nutritional Profile of Arancina SF

The above-mentioned description of *arancina* SF is naturally approximated when speaking of amounts concerning raw materials and ingredients. For this reason, and the nature of the TSG proposal, it is impossible to give a reliable description of the nutritional profile of this food.

However, a possible description for red and white *arancina* SF may be offered based on certain web references (Anonymous 2020a, b), taking into account that these data are only an indication and their reliability is extremely affected by different factors, including the typology (western or eastern version, etc.). Weight is naturally critical: once more, it has to be remembered that *arancina* SF may vary from 150 to 500 g! In our knowledge, there are not scientific papers concerning this topic.

Taking into account the above-mentioned premise, the following nutrition facts may be assumed for *arancina* (and *arancino…*) SF:

– Energy (as Kcal): 270 per 100 g (342 kcal are also reported, without relation to weights),
– Lipids: 12.01–19 g,
– Saturated fat: 4.5–6.0 g,
– Carbohydrates: 27.3 or 35 g,
– Sugars: 1.5 g,
– Fibres: 1.5 g,
– Protein: 11.3 g,
– Sodium: 80–383 mg,
– Cholesterol: 43–107 mg.

A reflection has to be made with reference to the main amount of all expressed nutrients: carbohydrates. This nutrient probably defines the main feature of western and eastern versions, with the additional presence of lipids (also found with degraded fatty acid chains…). Consequently, more research is surely needed in relation to *arancina* SF. In fact, some reported research has already investigated the role of SF when speaking of cardiovascular and other diseases. Because of the notable amount of lipids (especially with reference to degraded fatty acids) and the abundance of carbohydrates with respect to the sum of lipids, carbohydrates, and protein content (approximate value: 50.6%—the moisture amount is not far from 49%), it may be assumed that this food should be preferred a few times only per month (high caloric intake). On the other side, the consumption of these SF should be preferred if compared to typical fast-food products based on the 'weight' factor: one single *arancina* may weigh 150–200 g and abundantly cover main nutritional needs…

References

Alfiero S, Giudice AL, Bonadonna A (2017) Street food and innovation: the food truck phenomenon. Brit Food J 119(11):2462–2476. https://doi.org/10.1108/BFJ-03-2017-0179
Allison N (2018) Food, the city, and the street. Rev Adm Empresas 58(3):345
Alves da Silva S, Cardoso RCV, Góes JTW, Santos JN, Ramos FP, Bispo de Jesus R, Sabá do Vale R, Teles da Silva PS (2014) Street food on the coast of Salvador, Bahia, Brazil: a study from the socioeconomic and food safety perspectives. Food Control 40(1):78–84. https://doi.org/10.1016/j.foodcont.2013.11.022

Alves RMV, De Luca S, Grigoli CI, DenderAGF V, De Assis FF (1996) Stability of sliced Mozzarella cheese in modified-atmosphere packaging. J Food Prot 59(8):838–844. https://doi.org/10.4315/0362-028X-59.8.838

Andolina C (2019) The Panormus Street food: la rosticceria palermitana conquista i mercati inglesi. www.orogastronomico.it. Available https://www.orogastronomico.it/news-ed-eventi/the-panormus-street-food-la-rosticceria-palermitana-in-inghilterra/. Accessed 7 April 2020

Anonymous (2020a) Arancini di riso o arancine. Cibo360.it. Available https://www.cibo360.it/cucina/mondo/arancini_riso.htm. Accessed 10 April 2020

Anonymous (2020b) Database degli alimenti e contacalorie - Arancino. www.fatsecret.it. Available https://www.fatsecret.it/calorie-nutrizione/generico/arancino. Accessed 7 April 2020

Balsari P, Manzone M, Marucco P, Tamagnone M (2013) Evaluation of seed dressing dust dispersion from maize sowing machines. Crop Prot 51:19–23. https://doi.org/10.1016/j.cropro.2013.04.012

Bartolotta S (2010) Lengua y cultura gastronómica italianas. Universidad Nacional de Educación a Distancia (UNED) Editorial, Madrid

Basile G (2015) Piaceri e misteri dello street food palermitano. Storia, aneddoti e sapori del cibo di strada più buono d'Europa. Dario Flaccovio Editore, Palermo. ISBN 978-88-579-0461-0

Brown K (2018) Burek, Da! Sociality, context, and dialect in Macedonia and beyond. In: Montgomery DW (ed) Everyday life in the Balkans. Indiana University Press, Bloomington

Cabibbo S (2018a) Arancine al burro – ricetta originale siciliana. Available https://blog.giallozafferano.it/mastercheffa/arancine-al-burro/. Accessed 10 April 2020

Caldara J (2013) Il fenomeno dell'Italian sounding. Dissertation, University of Padua

Camilleri A (2018) Gli Arancini di Montalbano. Sellerio Editore, Palermo

Rubio C (2014) Unti e bisunti. Sperling & Kupfer, Segrate

Chukuezi CO (2010) Food safety and hyienic practices of street food vendors in Owerri, Nigeria. Stud Sociol Sci 1(1):50–57. https://doi.org/10.3968/j.sss.1923018420100101.005

CNC (2014a) Accordo tra Comune di Palermo e Consiglio Nazionale dei Chimici (CNC) per la partecipazione alla realizzazione di un sistema di salvaguardia e garanzia della tradizione gastronomica palermitana', Prot. 646/14/cnc/fta. Consiglio Nazionale dei Chimici (CNC), Rome. Available https://www.chimicifisici.it/wp-content/uploads/2018/10/20131210_accordo_firmato_dal_Presidente_del_CNC.pdf. Accessed 7 April 2020

CNC (2014b) DOMANDA DI REGISTRAZIONE DI UNA STG - Art. 8 - Regolamento (UE) n. 1151/2012 del Parlamento Europeo e del Consiglio del 21 novembre 2012 sui regimi di qualità dei prodotti agricoli e alimentari"ARANCINA". Annex to the document 'Accordo tra Comune di Palermo e Consiglio Nazionale dei Chimici (CNC) per la partecipazione alla realizzazione di un sistema di salvaguardia e garanzia della tradizione gastronomica palermitana', Prot. 646/14/cnc/fta. Consiglio Nazionale dei Chimici (CNC), Rome. Available https://www.chimicifisici.it/wp-content/uploads/2018/10/ARANCINA_CNC__STG_2014.pdf. Accessed 7 April 2020

Crowhurst DG, Creed PG (2001) Effect of cooking method and variety on the sensory quality of rice. Food Serv Technol 1(3):133–140. https://doi.org/10.1046/j.1471-5740.2001.d01-3.x

Cumbo EC (2000) La Festa Del Pane: food, devotion and ethnic identity, the feast of San Francesco Di Paola, Toronto. J Stud Food Soc 4(2):47–66. https://doi.org/10.2752/152897900786732781

D'Ignoti S (2019) The gender fight behind Sicily's most iconic snack. British Broadcasting Corporation (BBC), London. Available https://www.bbc.com/travel/story/20190415-the-gender-fight-behind-sicilys-most-iconic-snack. Accessed 10 April 2020

da Vico L, Biffi B, Agostini S, Brazzo S, Masini ML, Fattirolli F, Mannucci E (2010) Validation of the Italian version of the questionnaire on nutrition knowledge by Moynihan. Monaldi Arch Chest Dis 74(3):140–146. Available https://monaldi-archives.org/index.php/macd/article/download/263/252. Accessed 10 April 2020

Di Giorgi V (2017) Lo Spleen di Palermo. Rappresentazioni e identità intorno al cibo di strada palermitano. Dissertation, Università Ca'Foscari, Venice

Dittrich C (2017) Street food, food safety and sustainability in an emerging mega city: insights from an empirical study in Hyderabad, India. In Xaxa V, Saha D, Singha R (eds) Work, institutions and sustainable livelihood. Palgrave Macmillan, Singapore

Droga V, Lo Bue G (2009) L'Intervista - Lo storico Gaetano Basile: "Una palla di riso che resiste ai secoli". www.livesicilia.it. Available https://livesicilia.it/2009/12/13/lo-storico-gaetano-basile-una-palla-di-riso-che-resiste-ai-secoli_33648/. Accessed 10 April 2020

European Commission (2019) DOOR database. European Commission, Brussels. Available https://ec.europa.eu/agriculture/quality/door/list.html?recordStart=0&recordPerPage=10&recordEnd=10&sort.mileston. Accessed 7 April 2020

European Commission (2020) eAmbrosia—the EU geographical indications register. Eruopean Commission, Brussels. Available https://ec.europa.eu/info/food-farming-fisheries/food-safety-and-quality/certification/quality-labels/geographical-indications-register/. Accessed 7 April 2020

European Parliament and Council (2012) Regulation (EU) No 1151/2012 of the European Parliament and of the Council of 21 November 2012 on quality schemes for agricultural products and foodstuffs. OJ Eur Union L 343:1–29

FAO (2009) Good hygienic practices in the preparation and sale of street food in Africa. Food and Agriculture Organization of the United Nations (FAO), Rome. Available https://www.fao.org/3/a0740e/a0740e00.pdf. Accessed 09 April 2020

Fellows P, Hilmi M (2012) Selling street and snack foods. FAO Diversification booklet number 18. Rural Infrastructure and Agro-Industries Division, Food and Agriculture Organization of the United Nations (FAO), Rome. Available https://www.fao.org/docrep/015/i2474e/i2474e00.pdf. Accessed 7 April 2020

Gentilcore D (2010) Pomodoro!: a history of the tomato in Italy. Columbia University Press, New York, New York

Giannakourou MC, Taoukis PS (2003) Kinetic modelling of vitamin C loss in frozen green vegetables under variable storage conditions. Food Chem 83(1):33–41. https://doi.org/10.1016/S0308-8146(03)00033-5

Guzzo S (2014) Cross-cultural adaptation as a Form of translation: translating food in the UK Italian Community. Testi e linguaggi 8:161–169. Available https://elea.unisa.it/bitstream/handle/10556/3350/14_Guzzo.pdf?sequence=1&isAllowed=y. Accessed 10 April 2020

Gwinner T (2018) The origin and evolution of ragù bolognese (Bolognese Sauce). Int J Arts Humanit 4(3):16–28

Haddad MA, Parisi S (2020) Evolutive profiles of Mozzarella and vegan cheese during shelf-life. Dairy Industries International 85(3):36–38

Hansen M, Jakobsen HB, Christensen LP (2000) The aroma profile of frozen green peas used for cold or warm consumption. In: Proceedings of the 9th Weurman Flavour Research Symposium, Technical University of Munich, 22–25 June 1999, p 64

Hopping BN, Erber E, Mead E, Sheehy T, Roache C, Sharma S (2010) Socioeconomic indicators and frequency of traditional food, junk food, and fruit and vegetable consumption amongst Inuit adults in the Canadian Arctic. J Human Nutr Diet 23:51–58. https://doi.org/10.1111/j.1365-277X.2010.01100.x

Hung YC, Thompson DR (1989) Changes in texture of green peas during freezing and frozen storage. J Food Sci 54(1):96–101. https://doi.org/10.1111/j.1365-2621.1989.tb08576.x

Iannizzotto S (2016) Si dice arancino o arancina? Accademia della Crusca, Florence. Available https://accademiadellacrusca.it/it/consulenza/si-dice-arancino-o--arancina/1043. Accessed 10 April 2020

Johns J (2002) Arabic administration in Norman Sicily: the royal diwan. Cambridge University Press, Cambridge

Kraig B, Sen CT (eds) (2013) Street food around the world: an encyclopedia of food and culture. Abc-Clio, Santa Barbara

Lee R (2017) Changing Hong Kong street food culture: an analysis of the Hong Kong food truck pilot scheme. Henry Luce China - Environment Program, Occidental College, August 16, 2017

Long-Solís J (2007) A survey of street foods in Mexico city. Food Foodways 15(3–4):213–236. https://doi.org/10.1080/07409710701620136

Maiatsky M (2009) Pirojki à la bolognaise: Le miroir russe de l'université européenne. Multitudes 39(4):100–108. https://doi.org/10.3917/mult.039.0100

Maranzano B (2014) Lo sviluppo del fenomeno "street food": il cibo di strada a Palermo ieri e oggi. Dissertation, University of Pisa, Italy

Marino AMF, Giunta R, Salvaggio A, Farruggia E, Giuliano A, Corpina G (2010) Valutazione microbiologica di un prodotto alimentare tipico della tradizione siciliana: l'arancino. Riv Sci Aliment 39(4):17–21. Available https://fosan.it/system/files/Anno_39_4_3.pdf. Accessed 10 April 2020

Mele MN (2016) Arancinario. Il cuore croccante della Sicilia. Autopublished. ISBN-13: 979-1220016087

Ministero delle politiche agricole alimentari e forestali (2014) Quattordicesima revisione dell'elenco dei prodotti agroalimentari tradizionali. Ministero delle politiche agricole alimentari e forestali, Rome. Available https://www.politicheagricole.it/flex/cm/pages/ServeBLOB.php/L/IT/IDPagina/3276. Accessed 7 April 2020

Mohan CO, Remya S, Ravishankar CN, Vijayan PK, Srinivasa Gopal TK (2014) Effect of filling ingredient on the quality of canned yellowfin tuna (Thunnus albacares). Int J Food Sci Technol 49(6):1557–1564. https://doi.org/10.1111/ijfs.12452

Nessi F (2018) Marketing islamico: una sfida per le imprese italiane del settore food & beverage. Dissertation, Università Ca'Foscari, Venice

Pizzuto Antinoro G (2002) Gli Arabi in Sicilia e il modello irriguo della Conca d'Oro. Regione siciliana, Assessorato agricoltura e foreste, Palermo

Parente G (2007) Cibo veloce e cibo di strada. Le tradizioni artigianali del fast-food in Italia alla prova della globalizzazione. Storicamente 3, 2. https://doi.org/10.1473/stor389

Pensovecchio F (2017) Sicilia continente gastronomico: i grandi chef e la tradizione. Giunti Editore, Florence

Privitera D (2015) Street food as form of expression and socio-cultural differentiation. In: Proceedings of the 12th PASCAL international observatory conference, Catania

Ricci G (2014) Ispanismi nel "siciliano" di Andrea Camilleri. In: Toro LL, Luque R (eds) Léxico Español Actual IV. Libreria Editrice Cafoscarina, Venice, pp 173–191

Rosales S (2014) "This street is essentially Mexican": an oral history of the Mexican American Community of Saginaw, Michigan, 1920–1980. Mich Hist Rev 40(2):33–62. https://doi.org/10.5342/michhistrevi.40.2.0033

Silverman C (2012) Education, agency, and power among Macedonian Muslim Romani women in New York City. Signs J Women Cult Soc 38(1):30–36. https://doi.org/10.1086/665803

Simopoulus AP, Bhat RV (2000) Street foods. Karger AG, Basel

Sorensen JN, Edelenbos M, Wienberg L (2003) Drought effects on green pea texture and related physical-chemical properties at comparable maturity. J Am Soc Horticult Sci 128(1):128–135. https://doi.org/10.21273/JASHS.128.1.0128

Sozer N (2009) Rheological properties of rice pasta dough supplemented with proteins and gums. Food Hydrocoll 23(3):849–855. https://doi.org/10.1016/j.foodhyd.2008.03.016

Steyn NP, Labadarios D (2011) Street foods and fast foods: how much do South Africans of different ethnic groups consume? Ethnic Dis 21(4):462–466

Steyn NP, Mchiza Z, Hill J, Davids YD, Venter I, Hinrichsen E, Opperman M, Rumbelow J, Jacobs P (2013) Nutritional contribution of street foods to the diet of people in developing countries: a systematic review. Pub Health Nutr 17(6):1363–1374. https://doi.org/10.1017/S136898009001301158

Summerfield G (ed) (2009) Patois and linguistic Pastiche in modern literature. Cambridge Scholars Publishing, Cambridge

Suwannaporn P, Linnemann A (2008) Rice-eating quality among consumers in different rice grain preference countries. J Sens Stud 23(1):1–13. https://doi.org/10.1111/j.1745-459X.2007.00129.x

Tinker I (1999) Street foods into the 21st century. Agric Human Val 16:327–333

Toktassynova Z, Akbaba A (2017) Content analysis of on-line booking platform reviews over a restaurant: a case of pizza locale in Izmir. Avrasya Sosyal ve Ekonomi Araştırmaları Dergisi 5(5):242–249

Tomaiuolo S (2009) 'I am Montalbano/Montalbano sono': fluency and cultural difference in translating Andrea Camilleri's fiction. J Anglo-Ital Stud 10:201–219

Wright CA (1996) Cucina Arabo-Sicula and Maccharruni. Al-Masāq 9(1):151–177. https://doi.org/10.1080/09503119608577029

Chapter 3
Palermo's Street Foods. The Authentic *Sfincionello*

Abstract The so-called 'Street Food' is often claimed to represent a different version of the edible product in the Mediterranean Basin, in the same area of the 'Mediterranean Diet'. This style is perceived and claimed to be a synonym of 'safe', 'healthy', and 'hygienic' food eating behaviour. Actually, related foods are mainly considered 'safe' because of their composition and the presence of antioxidants. On the other side, 'street food' concerns only ready-to-eat foods and beverages which are prepared and sold literally in city and town streets and in similar places (including also small food trucks). This feature is observed worldwide (United Kingdom, United States of America, etc.) and can be studied in Italy, and especially in Sicily, where cheapness and other reasons have to be considered. The prominence of fried street foods in Sicily should be considered as a peculiar heritage of ancient civilisations, including *arancina/arancino* types and pizza (or *focaccia*)-like products. The present Chapter concerns one of the most known street food specialities in the Palermo area: the '*sfincionello*' product—with some digression concerning similar foods—by different viewpoints including history, possible 'authenticity' features, identification of raw materials, preparation procedures, concomitant alternative recipes, and nutrition facts.

Keywords Arabic cuisine · Gluten · Leavening · *Sfincione* · *Sfincionello* · Sicily · Street food

Abbreviations

CNC	Consiglio Nazionale dei Chimici
FAO	Food Agriculture Organization of the United Nations
FBO	Food business operator
HACCP	Hazard Analysis and Critical Control Points
MRP	Maillard reaction product
MD	Mediterranean Diet
MiPAAF	Ministero delle politiche agricole alimentari e forestali
OO	Olive oil

© The Author(s), under exclusive license to Springer Nature Switzerland AG 2020   43
M. Barone and A. Pellerito, *Sicilian Street Foods and Chemistry*,
Chemistry of Foods, https://doi.org/10.1007/978-3-030-55736-2_3

SF   Street food
TSG   Traditional speciality guaranteed

3.1   Are Sicilian Street Foods an Expression of Mediterranean Diet?

The so-called 'Street Food' is often claimed to represent a different version of Mediterranean foods (in the Mediterranean Basin, including North Africa and Middle East regions) (Barbieri et al. 2014; Chammem et al. 2018; David 2002; Delgado et al. 2017; Haddad et al. 2020a, b; Matalas et al. 2001). Could this interpretation be reliable?

With concern to Mediterranean Diet (MD), it is widely recognised that the basis of the MD pyramid (Bach-Faig et al. 2011; Davis et al. 2015; Willett et al. 1995) is composed of vegetable products and transformed foods from vegetable sources, while the consumption of dairy and meat foods is low (Keys 1995). Anyway, a recommended amount (or number of times) per week is one of the critical bases for MD pyramid (Delgado et al. 2017).

Also, MD is perceived and claimed to be a synonym of 'safe', 'healthy', and 'hygienic' food eating style, even if recommendations do not refer to chemical, physical, and microbiological food risks in the ambit of 'Hazard Analysis and Critical Control Points' (HACCP) evaluations. Actually, MD-related foods are mainly considered because of their composition (low calories; low carbohydrates, sugars, and lipids), and the presence of antioxidants including polyphenols (Haddad et al. 2020a, b; Parisi 2019, 2020).

With concern to 'street food' (SF), the related definition—according to the Food Agriculture Organization of the United Nations (FAO)—concerns only 'ready-to-eat' foods and beverages which are prepared and sold by different food business operators (FBO) without immobile location: production and selling services are literally in city and town streets and in similar places, including also small food trucks (Anenberg and Kung 2015; FAO 1997; Maranzano 2014; Pitte 1997; Vanschaik and Tuttle 2014). A small example is the 'Panormus Street Food' SF mobile caterer company currently working in the United Kingdom (Andolina 2019). Other well-studied situations concern the 'take away' services in New York City or Los Angeles, of food truck services in Italy (Alfiero et al. 2017, 2018; Bhimji 2010; Codesal 2010; Messer and Cohen 2007; Shackman et al. 2015).

SF is diffused as a cultural heritage in all known urbanised areas of the world, suggesting a clear correlation between social aggregation and food consumption (Bell and Loukaitou-Sideris 2014; Booth and Coveney 2007; Chong and Eun 1992; Cortese et al. 2016; de Suremain 2016; Heinzelmann 2014; Helou 2006; Kollnig 2020; Larcher and Camerer 2015; Manning 2009; Maranzano 2014; McHugh 2015; Morton 2014; Patel et al. 2014; Pilcher 2017; Sheen 2010; Simopoulus and Bhat 2000; Steven 2018; Webb and Hyatt 1988; Wilkins and Hill 2009). Moreover, SF is the synonym of historical dominations, business strategies, and cultural (ethnic) differences. Non-technical disciplines such as history, architectural design, marketing, and other ambits without relation to chemistry at least can be comprised in this context. On the other side, SF or MD-related foods have surely several chemical, microbiological, and technological features which can explain the 'success' and the temporal persistence of certain traditions (Parisi and Luo 2018; Parisi et al. 2004, 2019; Singla et al. 2018). This reflection has a peculiar importance in certain cultural areas such as Italy and especially Sicily, where cheapness and other reasons have to be considered when speaking of comparison between 'fast-food' products on the one hand, and typical SF products on the other side (Parente 2007). In addition, the consumption of SF is generally associated with a social elevation in terms of cultural defense of historical traditions (Di Giorgi 2017; Parente 2007).

The fragmentation of the current food and beverage market has been reported many times in the recent literature (Alves da Silva et al. 2014; FAO 2009; Hopping et al. 2010; Long-Solís 2007; Maranzano 2014; Privitera 2015; Simopoulus and Bhat 2000; Steyn and Labadarios 2011; Steyn et al. 2013; Tinker 1999). Consequently, many regional or national dishes have spread across the world with partially predictable results such as mixtures between Italian *pizza* and *kebab*, the Russian-style *pirozhki* dish with *bolognese* sauce (Maiatsky 2009) or the Balkan-style food named '*pizza/burek*' (Brown 2018; Kraig and Sen 2013; Silverman 2012). This phenomenon of cultural 'contamination' (Toktassynova and Akbaba 2017) could also damage or modify certain traditions, in Sicily and worldwide. Many and many specialities in Sicily should be discussed, from hard-boiled octopus, *cicireddu* (fried little fish), and fried *supplì* (rice croquettes filled with tomatoes, mozzarella cheese, and meat) (Parente 2007). However, the argument is extremely broad, and it should be evaluated critically in more than one book.

The prominence of fried SF in Sicily, and in the Palermo area, in particular, should be considered as a peculiar heritage of ancient civilisations, above all the Arabic tradition (Di Giorgi 2017; Parente 2007). These foods, including *arancina/arancino* (Chap. 2) or pizza (or *focaccia*)-like products, have some typical features in the capital city of Sicily, Palermo.

The present Chapter concerns one of the most known SF specialities in Siciliy, and in the Palermo area: the '*sfincionello*' SF.

3.2 A Typical Palermo's Food: The *Sfincionello*

The SF speciality commonly named '*sfincionello*' is found only in the Sicilian capital city, Palermo, on the western side. This SF has a typical name with several similarities if compared with other foods, and also an alternative version in the same area. This and other similar foods may be defined as a cultural heritage of the Arabic cuisine (Barbagli and Barzini 2010; CNC 2014a, b).

In detail, and as explained in the following Sections, it is based on a peculiar 'pie 'named '*pane pizza*' (pizza bread) in the Italian language. Its texture and appearance remember a small sponge, and this appearance justifies the '*sfincionello*' name (CNC 2014b).

According to reported references, *sfincionello* is one of three products with a similar name and some shared features (CNC 2014b):

(a) The *sfincione* (normally considered as the typical SF in Palermo),
(b) The *sfincionello* (assumed to be a little version of *sfincione*),
(c) The *sfincia di San Giuseppe* (a sweet food, completely different from *sfincione* and *sfincionello*, meaning 'The Saint Joseph's Sponge' in Italian language),
(d) The *sfincetta*—another sweet food which should be the real original recipe for *sfincione* in spite of its sweet features.

Before proceeding with the discussion for *sfincionello*, a little premise has to be made in relation to the origin of these foods and similarities in the Arabic and European cuisine.

3.2.1 Sfincione, Sfincionello, Sfincetta, and Sfincia di San Giuseppe. Common Origins

As above explained, *sfincione, sfincionello, sfincia di San Giuseppe,* and *sfincetta* compose an ideal food 'quadruple' with common features. In detail:

(a) The *sfincione* corresponds to a soft and leavened sponge-*like pane pizza* pie with the following topping ingredients: tomato, anchovy, oregano, onion, and pieces of *caciocavallo ragusano* (a typical Sicilian cheese) (Gambino 2002; Pasqualone et al. 2011; Pignataro 2018).
(b) The *sfincionello* SF is similar to *sfincione* because of its nature and its reduced dimensions. After all, the name would mean 'little *sfincione*' in Italian language. The basic (round or oval) *pane pizza* pie (obtained with wheat flour 00) is superficially coated with (reconstituted) double concentrate of tomato, olive oil, onion, breadcrumbs, sea salt, and dried oregano (CNC 2014b).
(c) The *sfincetta* appears to be the original base for *sfincione*, in spite of its sweet food nature. Substantially, it has the same nature of *sfincione* except for the presence of sugar cane as coating.
(d) Finally, the *sfincia di San Giuseppe* is a peculiar pancake with completely different features (it is a sweet food!). However, its place in this ideal 'triad' is needed because of common origins.

Two of these foods with the exclusion of *sfincionello* and *sfincette* have been recognised as Italian traditional food products or 'prodotti agroalimentari tradizionali italiani' (PAT) of the Italian Ministry of Agricultural, Food and Forestry Policies (MiPAAF) (Ministero delle politiche agricole alimentari e forestali 2014). On the other side, in 2014, a project concerning the application for registration of a 'traditional speciality guaranteed' (TSG), according to the Title III of the Regulation (EU) No. 1151/2012 of the European Parliament and of the Council of 21 November

2012 on quality schemes for agricultural products and foodstuffs (European Parliament and Council 2012) has been proposed in Italy with relation to *sfincionello* (as different from *sfincione*) (CNC 2014a, b). This proposal followed the creation of a '*City Brand Panormvs*' for the protection of Palermo products (Maranzano 2014). The TSG proposal aims at the creation of the '*Panormvs—street food*' brand specifically linked to the TGS product(s) and the realisation of an international SF festival in Palermo (CNC 2014a). At present, there is a 'Panormus Street Food' SF mobile caterer company currently working in the United Kingdom (Andolina 2019). On the other side, there is not available information concerning the *sfincionello* TSG: there are not Italian applications in this ambit as TSG on 24 March 2020 (European Commission 2019, 2020).

In general, all these foods (only two of these specialities are SF, while sweet foods do not belong to the ambit of street food vendors) have a similar name from Latin *spongia* (sponge), the greek σπόγγος (sponge), and/or from the Arabic term جفنسا (*isfanğ* or sponge) (CNC 2014b). It may be assumed that all these foods are a new local version of certain Arabic traditional recipes. As a single example, the sweet food is believed to be found with different names and attributions (as sweet fried pancakes) both in the Holy Bible and in the Koran.

Analogous leavened (with *Saccharomyces cerevisiae* yeasts) versions of the same heritage may be found in Spain, particularly in the Andalusia region, where *churros* (Bermudo et al. 2006; Castro-Lopez et al. 2017; Mir-Bel et al. 2013) are normally found (from Roman Empire recipes, probably), or in Morocco, where a '*beghrir*' pancake (similar to *churros*) is found (Charbonnier 2013; CNC 2014b; Palladino 2018; Salloum et al. 2013). Interestingly, Palermo *sfincette* are similar to both *churros* and *beghrir* foods.

Basically, this SF is a pizza-like product coated with different ingredients (Barbagli and Barzini 2010; CNC 2014b). In detail, the product may be described with the following attributions (CNC 2014c; Marino et al. 2010):

(1)   A ready-to-eat food, of a circular shape, with an approximate diameter or maximum length: 30 cm, and weight approximately equal to 700 g
(2)   The food may have to show a visible yellowish rind around the circular area.

The coating is composed of a light-red tomato sauce (including also four or five cooked tomato wedges) and with the following colorimetric attribution: Cyan: 0%; Magenta: 73%; Yellow: 100%; and Black: 0%. Also, the following ingredients may be used: onions, anchovies, and *caciocavallo* cheese (Bonsaver et al. 2019; Buscemi et al. 2011; Di Giorgi 2017; Parente 2007; Pasqualone et al. 2011; Rajendran 2004).

An alternative version of the same *sfincionello* SF can be elaborated with an ellipsoidal (oval) shape with axes of 12 and 30 cm (Fig. 3.1). In this situation, the weight would be lower than the first version (approximately 300 g) and the coating is obtained with the same light-red tomato sauce.

Anyway, both *sfincionello* versions have a well-known feature in common: the pie has to show a high, soft, and inner structure which could be defined as a 'honeycomb-like' network. The typical compressible and 'fluffy' structure is obtained because of

Fig. 3.1 The *sfincionello* SF is a pizza-like product coated with different ingredients (Barbagli and Barzini 2010; CNC 2014b, c). The product may be realised in two versions. One of these is shown here: the *sfincionello* SF can be elaborated with an ellipsoidal (oval) shape

the peculiar leavening method: the process allows the production of pies with low density, similarly to the similar *sfincione* (Billitteri 2002; CNC 2014b; Parente 2007).

3.3   *Sfincionello* SF. Ingredients and Preparation

3.3.1   *Raw Ingredients*

Basically, the preparation of *sfincionello* types concerns these ingredients:

(a) W*heat flour 00 (double zero)* (Balsari et al. 2013),
(b) *Water,*
(c) *Yeasts,*
(d) Double tomato concentrate (to be reconstituted),
(e) Olive oil (OO),
(f) Onions,
(g) Breadcrumbs,
(h) Sea salt,

(i) Dry oregano.

The above-mentioned list does not report anchovies and cheeses, meaning that the TSG proposal does not consider these ingredients as mandatory (while red tomato and related wedges are mandatory, despite their introduction in the Sicilian tradition after 1492 A.D., some centuries after the Arabic domination).

Interestingly, and differently from other Palermo's SF, some ingredients can be used without precise specifications. The example of OO is peculiar, because the production of *arancina* or *arancino* foods (Chap. 1) requires extra-virgin olive oil (EVOO) (Bianchi et al. 2001; Cicerale et al. 2012; Oliveras-López et al. 2013), clearly stating that authenticity is important. On the other hand, this requirement is not requested in the present ambit, perhaps explaining partially the cheapness of *sfincionello* SF.

Another reflection concerns the characterisation of wheat flour 00 (because its type is critical when speaking of the leavening process). This ingredient has to contain dry gluten between 9.5 and 11%, and protein content between 9 and 12.5%. Anyway, protein should be 9.0% as a minimum amount, while maximum recommended moisture should not exceed 14.5% (Matz 1991). These data appear critical in this ambit, and they can also explain differences in terms of elasticity and toughness of flour gluten. The related indicator is named 'W indicator' (Kohajdová et al.2013; Pshenichnikova et al. 2008; Stoica et al. 2013). In this ambit, W should be recommended between 220 and 380 (medium-strength to strong flour). In these conditions, the higher the protein content, the higher the W indicator, and consequently the longer the rising (leavening) time, as recommended for *sfincionello*. Should the wheat flour have these features, the produced carbon dioxide would be easily trapped during fermentation by means of a gluten network inside pies. This process explains the honeycomb-like structure of this food (Fig. 3.2). Another (chemical and historical) reflection should concern the peculiar texture of *sfincionello* and similar products because of the availability of simple sugars. At present, this availability is substantially dependent on the quality of wheat flour 00, while artisanal recipes were probably followed by SF vendors with the use of boiled potatoes (potato starch surely contains a most promising amount of sugars) (Kohyama and Nishinari 1991; KoVan Den et al. 1986). Currently, boiled potatoes are used for homemade *sfincione* products only.

3.3.2 Preparation and Storage

3.3.2.1 Dough

The dough is prepared with wheat flour 00, water, sea salt, and brewer's yeast (*S. cerevisiae*). The mixture should be obtained with (CNC 2014b):

(a) Water, 1000 g,
(b) Sea salt, 20–25 g,
(c) Wheat flour 00, 100 g,

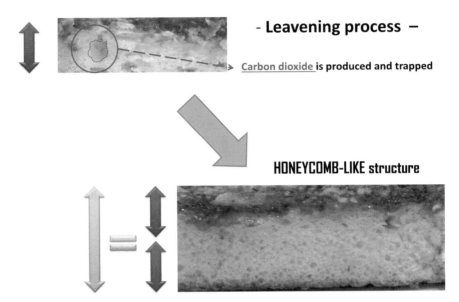

Fig. 3.2 The leavening process is critical when speaking of *sfincionello* SF. The higher the protein content in flour, the higher the W indicator (recommended between 220 and 280), and consequently the longer the rising (leavening) time. Should the wheat flour have these features, the produced carbon dioxide would be easily trapped during fermentation by means of a gluten network inside pies. This process explains the honeycomb-like structure of this food and the related height (approximately two times the original height before leavening)

(d) Brewer's yeast, 50 g in (additional) warm water, 100 g.

The mixture has to give a homogeneous and elastic enough mass. Subsequently, wheat flour 00 has to be added again (900 g) with the aim of obtaining a 'fluffy' and high layer structure. The duration of the whole process should be approximately 20 min (at low speed, temperature: 25–30 °C, environmental humidity 70–80%). The desired result is achieved on condition that the aqueous yield (amount of absorbed water by flour) is 50–100% if compared with used flour (in other terms: 500–1000 g of water per 1000 g of flour).

3.3.2.2 Leavening, Forming, and Coating

The prepared dough should be subdivided into a a series of semicircular masses (approximate weight: 500 g) placed for 30 min under a damp cloth. The aim of this procedure is to avoid that superficial moisture is not released excessively; otherwise, the worked dough would easily produce a sort of bread crust in contact with the normal (or moist) atmosphere. This process is the so-called 'leavening' step with the needed help of brewer's yeast (*S. cerevisiae*).

After 30 min, the forming step has to be carried out, and the TSG proposal explicitly affirms that the process has to be manually performed without automatic or semi-automatic tools or equipment. The same recommended procedure also explains how the forming of circular and ellipsoidal shapes from original semicircular masses should be carried out. Two circular pieces are realised (diameter: 30 cm, weigh: 500 g) and a third ellipsoidal piece (axes: 12 and 30 cm) is also obtained from the original dough. Substantially, the procedure means that one single 1500-to-2000 g dough should give two 500 g circular pieces and one ellipsoidal piece with the remaining amount. This result is the function of the dough yield, as above explained. Additionally, the manual forming has the aim of obtaining pieces with a peculiar external 'edge' of about 2 cm around the pies. This edge can give an idea of the height rise of the final *sfincionello* SF, and also explain similarities with often mentioned 'sponges' (Fig. 3.2).

After this step, prepared pies are covered superficially with above-described ingredients, and each prepared piece can be placed on dedicated baking pans into electric or wood ovens for an additional 60 min. It has to be considered that this process is not a cooking method but an additional leaving procedure aiming at helping the rise of *sfincionello* layers (the honeycomb-like network). The procedure tacitly considers an oven temperature of 500 °C, and the affirmation that this process is a simple leavening method may be questionable. However, it has to be also noted that baking pans for partially leavened pieces should be superimposed into ovens, meaning probably that each leavening pie is partially covered and protected from loss of superficial moisture (and carbon dioxide). Anyway, the procedure aims at

(a) Eliminating virtually all simple sugars,
(b) Producing the maximum amount of carbon dioxide,
(c) Obtaining a 'fluffy' and high-layered structure with exceptional features such as the ability to be compressed and to return to the initial state after the removal of external pressure.

Should a peculiar testing method be created in the future with relation to specific *sfincionello* properties, the above-mentioned pressure test might be taken into consideration.

With relation to the preparation of superficial coating (tomato sauce, tomato wedges, onions, breadcrumbs, etc.), the interested Reader is invited to consult the selected reference. This procedure has to be considered as a separate preparation.

3.3.2.3 Cooking

The final cooking has to be performed at 500 °C, as above anticipated. This process is critical because brewer's yeast (*S. cerevisiae*) has to be inactivated. The superficial coating has the function of a 'trapping' network for moisture and above all carbon dioxide (with the production and rise of the characteristic honeycomb-like network). In this way, pH is substantially low and residual yeasts and moulds are practically inactivated.

Recommended processing parameters should be

(a) Set oven temperature: 450 °C,
(b) Cooking temperature: 500 °C (it should be higher than the set oven temperature),
(c) Time: 3 min,
(d) Product temperature: inner product 75–80 °C, external coating or tomato 90–100 °C.

In these conditions, the pie should be completely cooked with a yellowish superficial colour, strong aroma, and above all—the absence of visible brown areas as a demonstration of Maillard reaction products (MRP). In relation to the last point, the reduced cooking time should explain the limited extension of MRP in this product (CNC 2014b; Parisi and Luo 2018; Parisi et al. 2019; Singla et al. 2018). Tomato sauce and wedges should show a good (semi-hard) texture because of partial aqueous removal from surfaces (at 90–100 °C).

3.3.2.4 Storage and Recommendations

There is only a recommendation for traditional *sfincionello* SF, according to the TSG proposal, when speaking of storage: this period should be limited to 24 h only, and possible preservation methods based on freezing, similar options, or the simple under vacuum packaging have to be avoided. Consequently, this product is a real SF food. In addition, its shelf life does not exceed 24 h (CNC 2014b). Actually, there are not many available studies concerning preserved *sfincionello* SF at present (Basile 2015; Basile and Ardizzone 2019).

With reference to served recommendations, the *sfincionello* should be re-heated at 50 °C, as SF vendors in Palermo are accustomed to carrying out their service, and coated with a little OO amount with dry oregano pieces (CNC 2014b). Once more, there are not recommendations concerning the quality and authenticity of olive oils, probably meaning that authenticity is a matter of product typology instead of separated ingredients.

3.3.2.5 Other Variations

The above-mentioned recommendations have been identified as the key to authentic *sfincionello* SF so far (CNC 2014b). Actually, different references may report similar or different recipes (Cardella 2008; Myfitnesspal 2020).

In brief, these modifications concern the admitted ingredients, mentioning also 'double-milled' wheat flour (in Italian: '*farina rimacinata*') and wheat flour 0 as 1:1, respectively (total amount: 1000 g), cheese (*caciocavallo* type, 600 g per 1000 g of flour), brewer's yeast (20 g per 1000 g of wheat), and salt (20 g per 1000 g of wheat).

3.4 Nutritional Profile of *Sfincionello* SF

The above-mentioned description of *sfincionello* SF is naturally approximated when speaking of amounts concerning raw materials and ingredients. For this reason, and the nature of the TSG proposal among other folk descriptions, it is impossible to give a reliable description of the nutritional profile of this food.

However, a possible description for this SF could be offered for *sfincione* (it is a different food) based on certain web references (Anonymous 2020a, b), taking into account that these data are only an indication and their reliability is extremely affected by different factors. Weight cannot be considered as a critical factor because this food should not exceed 500 g as weight, while critical ingredients—wheat flour 00 and brewer's yeast above all—can determine important modifications based on their chemical properties and their typology, respectively. In addition, there are not scientific papers concerning these topics and also durability (it should not exceed 24 h, but only as a matter of historical habit), in our knowledge.

Taking into account the above-mentioned premise, the following nutrition facts may be assumed for *sfincione* SF:

- Energy (as kcal): 276 per 100 g (142 or 218 kcal are also reported),
- Lipids: 5–10 g,
- Saturated fat: 0 g,
- Carbohydrates: 25–27.1 g,
- Sugars: 2.6 g,
- Fibres: 1.0 g
- Protein: 5.9 g,
- Sodium: 300–396 mg,
- Cholesterol: 0 mg.

A reflection has to be made with reference to the main amount of all expressed nutrients: carbohydrates. This nutrient probably defines the main feature of *sfincione* SF, while the amount of sodium (300–396 mg per 100 g in *sfincione* against 232 mg per 100 g in *arancina* SF, Chap. 1) should be considered when speaking of hypertension and other cardiovascular diseases. Anyway, more research is surely needed in relation to *sfincionello* SF because its nutritional data should be different from *sfincione*, and there is not sufficient scientific literature in this ambit at present. It has to be considered also that the energy intake is similar to *arancina* SF, probably meaning that both foods are a good expression of Sicilian SF at the same time.

References

Alfiero S, Giudice AL, Bonadonna A (2017) Street food and innovation: the food truck phenomenon. Brit Food J 119(11):2462–2476. https://doi.org/10.1108/BFJ-03-2017-0179
Alfiero S, Wade B, Bonadonna A, Taliano A (2018) Defining the food truck phenomenon in Italy: a feasible explanation. In: Cantino V, Culasso F, Racca G (eds) Smart tourism. McGraw-Hill

Education S.r.l., Milan, pp 365–385. Available https://iris.unito.it/handle/2318/1664358#.Xo7 eJTdR3IU. Accessed 9 April 2020

Alves da Silva S, Cardoso RCV, Góes JTW, Santos JN, Ramos FP, Bispo de Jesus R, Sabá do Vale R, Teles da Silva PS (2014) Street food on the coast of Salvador, Bahia, Brazil: a study from the socioeconomic and food safety perspectives. Food Control 40(1):78–84. https://doi.org/10.1016/j.foodcont.2013.11.022

Andolina C (2019) The Panormus street food: la rosticceria palermitana conquista i mercati inglesi. www.orogastronomico.it. Available https://www.orogastronomico.it/news-ed-eventi/the-panormus-street-food-la-rosticceria-palermitana-in-inghilterra/. Accessed 7 April 2020

Anenberg E, Kung E (2015) Information technology and product variety in the city: the case of food trucks. J Urban Econ 90:60–78. https://doi.org/10.1016/j.jue.2015.09.006

Anonymous (2020a) Database degli alimenti e contacalorie – Sfincione. www.fatsecret.it. Available https://www.fatsecret.it/calorie-nutrizione/generico/sfincione. Accessed 8 April 2020

Anonymous (2020b) Sfincione palermitano. www.fornellidisicilia.it. Available https://www.fornel lidisicilia.it/ricetta/sfincione-palermitano/. Accessed 8 April 2020

Bach-Faig A, Berry EM, Lairon D, Reguant J, Trichopoulou A, Dernini S, Medina FX, Battino M, Belahsen R, Miranda G, Serra-Majem L (2011) Mediterranean diet pyramid today. Science and cultural updates. Pub Health Nutr 14(12A):2274–2284. https://doi.org/10.1017/S13689800110 02515

Balsari P, Manzone M, Marucco P, Tamagnone M (2013) Evaluation of seed dressing dust dispersion from maize sowing machines. Crop Prot 51:19–23. https://doi.org/10.1016/j.cropro.2013.04.012

Barbagli A, Barzini S (2010) Pane, pizze e focacce. Giunti Editore, Florence

Barbieri G, Barone C, Bhagat A, Caruso G, Conley ZR, Parisi S (2014) The influence of chemistry on new foods and traditional products. Springer International Publishing, Cham

Basile G (2015) Piaceri e misteri dello street food palermitano. Storia, aneddoti e sapori del cibo di strada più buono d'Europa. Dario Flaccovio Editore, Palermo. ISBN 978-88-579-0461-0

Basile G, Ardizzone A (2019) Frugando tra i mercati di Palermo. Una foto, una storia. Edizioni d'arte Kalós, Palermo

Bell JS, Loukaitou-Sideris A (2014) Sidewalk informality: an examination of street vending regulation in China. Int Plan Stud 19(3–4):221–243. https://doi.org/10.1080/13563475.2014. 880333

Bermudo E, Moyano E, Puignou L, Galceran MT (2006) Determination of acrylamide in foodstuffs by liquid chromatography ion-trap tandem mass-spectrometry using an improved clean-up procedure. Anal Chim Acta 559(2):207–214. https://doi.org/10.1016/j.aca.2005.12.003

Bhimji F (2010) Struggles, urban citizenship, and belonging: the experience of undocumented street vendors and food truck owners in Los Angeles. Urb Anthropol Stud Cultural Sys World Econ Dev 39(4):455–492

Bianchi G, Giansante L, Shaw A, Kell DB (2001) Chemometric criteria for the characterisation of Italian protected denomination of origin (DOP) olive oils from their metabolic profiles. Eur J Lipid Sci Technol 103(3):141–150. https://doi.org/10.1002/1438-9312(200103)103:3%3c141:: AID-EJLT141%3e3.0.CO;2-X

Billitteri D (2002) Homo Panormitanus. Cronaca di un'estinzione impossibile. Pietro Vittorietti Edizioni, Palermo

Bonsaver G, Reza M, Carlucci A (2019) The dynamics of cultural change: a theoretical frame with reference to Italy-USA relations. '900 Transnazionale 3(1):107–130. https://doi.org/10.13133/ 2532-1994_3.9_2019

Booth SL, Coveney J (2007) Survival on the streets: prosocial and moral behaviors among food insecure homeless youth in Adelaide, South Australia. J Hunger Environ Nutr 2(1):41–53. https:// doi.org/10.1080/19320240802080874

Brown K (2018) Burek, Da! Sociality, context, and dialect in Macedonia and beyond. In: Montgomery DW (ed) Everyday life in the Balkans. Indiana University Press, Bloomington

Buscemi S, Barile A, Maniaci V, Batsis JA, Mattina A, Verga S (2011) Characterization of street food consumption in Palermo: possible effects on health. Nutr J 10(1):119. https://doi.org/10.1186/1475-2891-10-119

Cardella G (2008) Sfincione palermitano. https://www.ricettedisicilia.net/. Available https://www.ricettedisicilia.net/piatti-unici/lo-sfincione-la-vera-pizza-dei-palermitani/. Accessed 10 April 2020

Castro-Lopez R, Gómez-Salazar JA, Cerón García A, Sosa-Morales ME (2017) Stability of palm olein oil with different antioxidants during repeated frying of churros (Spanish fried dough pastry). The Canadian Society for Bioengineering, Paper No. CSBE17101. Available https://www.csbe-scgab.ca/docs/meetings/2017/CSBE17101.pdf. Accessed 9 April 2020

Chammem N, Issaoui M, De Almeida AID, Delgado AM (2018) Food crises and food safety incidents in European Union, United States, and Maghreb Area: current risk communication strategies and new approaches. J AOAC Int 101(4):923–938. https://doi.org/10.5740/jaoacint.17-0446

Charbonnier A (2013) Réflexions sur la communication à propos des patrimoines en contexte de trekking dans le Haut-Atlas (Maroc). Collection EDYTEM. Cahiers de géographie 14(1):67–78. Available https://www.persee.fr/doc/edyte_1762-4304_2013_num_14_1_1225. Accessed 10 April 2020

Chong HK, Eun VLN (1992) 4 backlanes as contested regions: construction and control of physical. In: Huat CB, Edwards N (eds) Public space: design, use and management. Singapore University Press, Singapore

Cicerale S, Lucas LJ, Keast RSJ (2012) Antimicrobial, antioxidant and anti-inflammatory phenolic activities in extra virgin olive oil. Curr Opin Biotechnol 23(2):129–135. https://doi.org/10.1016/j.copbio.2011.09.006

CNC (2014a) Accordo tra Comune di Palermo e Consiglio Nazionale dei Chimici (CNC) per la partecipazione alla realizzazione di un sistema di salvaguardia e garanzia della tradizione gastronomica palermitana', Prot. 646/14/cnc/fta. Consiglio Nazionale dei Chimici (CNC), Rome. Available https://www.chimicifisici.it/wp-content/uploads/2018/10/20131210_accordo_firmato_dal_Presidente_del_CNC.pdf. Accessed 7 April 2020

CNC (2014b) DOMANDA DI REGISTRAZIONE - Art. 8 - Regolamento (UE) n. 1151/2012 del Parlamento Europeo e del Consiglio del 21 novembre 2012 sui regimi di qualità dei prodotti agricoli e alimentari - "SFINCIONELLO". Annex to the document 'Accordo tra Comune di Palermo e Consiglio Nazionale dei Chimici (CNC) per la partecipazione alla realizzazione di un sistema di salvaguardia e garanzia della tradizione gastronomica palermitana', Prot. 646/14/cnc/fta. Consiglio Nazionale dei Chimici (CNC), Rome. Available https://www.chimicifisici.it/wp-content/uploads/2018/10/SFINCIONELLO_CNC_STG_2014.pdf. Accessed 7 April 2020

CNC (2014c) DOMANDA DI REGISTRAZIONE - Art. 8 - Regolamento (UE) n. 1151/2012 del Parlamento Europeo e del Consiglio del 21 novembre 2012 sui regimi di qualità dei prodotti agricoli e alimentari - " "ARANCINA". Annex to the document 'Accordo tra Comune di Palermo e Consiglio Nazionale dei Chimici (CNC) per la partecipazione alla realizzazione di un sistema di salvaguardia e garanzia della tradizione gastronomica palermitana', Prot. 646/14/cnc/fta. Consiglio Nazionale dei Chimici (CNC), Rome. Available https://www.chimicifisici.it/wp-content/uploads/2018/10/ARANCINA_CNC__STG_2014.pdf. Accessed 8 April 2020

Codesal DM (2010) Eating abroad, remembering (at) home. Three foodscapes of Ecuadorian migration in New York, London and Santander. Anthropol Food 7. https://doi.org/10.4000/aof.6642

Cortese RDM, Veiros MB, Feldman C, Cavalli SB (2016) Food safety and hygiene practices of vendors during the chain of street food production in Florianopolis, Brazil: a cross-sectional study. Food Control 62:178–186. https://doi.org/10.1016/j.foodcont.2015.10.027

David E (2002) A book of Mediterranean food. New York Review of Books Inc., New York

Davis C, Bryan J, Hodgson J, Murphy K (2015) Definition of the Mediterranean diet; a literature review. Nutrients 7(11):9139–9153. https://doi.org/10.3390/nu7115459

de Suremain CÉ (2016) The never-ending reinvention of 'traditional food'. In: Sébastia B (ed) Eating traditional food: politics, identity and practices. Routledge, Abingdon

Delgado AM, Almeida MDV, Parisi S (2017) Chemistry of the Mediterranean diet. Springer International Publishing, Cham. https://doi.org/10.1007/978-3-319-29370-7

Di Giorgi V (2017) Lo Spleen di Palermo. Rappresentazioni e identità intorno al cibo di strada palermitano. Dissertation, Università Ca'Foscari, Venice

European Commission (2019) DOOR database. European Commission, Brussels. Available https://ec.europa.eu/agriculture/quality/door/list.html?recordStart=0&recordPerPage=10&recordEnd=10&sort.mileston. Accessed 07 April 2020

European Commission (2020) eAmbrosia—the EU geographical indications register. Eruopean Commission, Brussels. Available https://ec.europa.eu/info/food-farming-fisheries/food-safety-and-quality/certification/quality-labels/geographical-indications-register/. Accessed 7 April 2020

European Parliament and Council (2012) Regulation (EU) No 1151/2012 of the European Parliament and of the Council of 21 November 2012 on quality schemes for agricultural products and foodstuffs. OJ Eur Union L 343:1–29

FAO (1997) Street foods. FAO Food and Nutrition Paper N. 63. Report of an FAO Technical Meeting on Street Foods, Calcutta, India, 6–9 November 1995. Food and Agriculture Organization of the United Nations (FAO). Available https://www.fao.org/3/W4128T/W4128T00.htm/. Accessed 7 April 2020

FAO (2009) Good hygienic practices in the preparation and sale of street food in Africa. Food and Agriculture Organization of the United Nations (FAO), Rome. Available https://www.fao.org/3/a0740e/a0740e00.pdf. Accessed 9 April 2020

Gambino D (2002) Une nuit à Palerme. La pensee de midi 8(2):62–67. Available https://www.cairn.info/revue-la-pensee-de-midi-2002-2-page-62.htm#. Accessed 9 April 2020

Haddad MA, Abu-Romman S, Obeidat M, Iommi C, El-Qudah J, Al-Bakheti A, Awaisheh S, Jaradat DMM (2020a) Phenolics in Mediterranean and Middle East important fruits. J AOAC Int, in press

Haddad MA, Obeidat M, Al-Abbadi A, Shatnawi MA, Al-Shadaideh A, Al-Mazra'awi M, I, Iommi C, Dmour H, Al-Khazaleh JM (2020b) Herbs and medicinal plants in Jordan. J AOAC Int, in press

Heinzelmann U (2014) Beyond bratwurst: a history of food in Germany. Reaktion Books Ltd., London

Helou A (2006) Mediterranean street food: stories, soups, snacks, sandwiches, barbecues, sweets, and more from Europe, North Africa, and the Middle East. Harper Collins Publishers, New York

Hopping BN, Erber E, Mead E, Sheehy T, Roache C, Sharma S (2010) Socioeconomic indicators and frequency of traditional food, junk food, and fruit and vegetable consumption amongst Inuit adults in the Canadian Arctic. J Human Nutr Diet 23:51–58. https://doi.org/10.1111/j.1365-277X.2010.01100.x

Keys A (1995) Mediterranean diet and public health: personal reflections. Am J Clin Nutr 61(6):1321S-1323S. https://doi.org/10.1093/ajcn/61.6.1321S

Kohajdová Z, Karovičová J, Magala M (2013) Rheological and qualitative characteristics of pea flour incorporated cracker biscuits. Croat J Food Sci Technol 5(1):11–17

Kohyama K, Nishinari K (1991) Effect of soluble sugars on gelatinization and retrogradation of sweet potato starch. J Agric Food Chem 39(8):1406–1410. https://doi.org/10.1021/jf00008a010

Kollnig S (2020) The 'good people' of Cochabamba city: ethnicity and race in Bolivian middle-class food culture. Lat Am Caribb Ethn Stud 15(1):23–43. https://doi.org/10.1080/17442222.2020.1691795

Den KoVan T, Biermann CJ, Marlett JA (1986) Simple sugars, oligosaccharides and starch concentrations in raw and cooked sweet potato. J Agric Food Chem 34(3):421–425. https://doi.org/10.1021/jf00069a010

Kraig B, Sen CT (eds) (2013) Street food around the world: an Encyclopedia of food and culture. Abc-Clio, Santa Barbara

Larcher C, Camerer S (2015) Street food. Temes de Dissen 31:70–83. Available https://core.ac.uk/download/pdf/39016176.pdf. Accessed 9 April 2020

Long-Solís J (2007) A survey of street foods in Mexico city. Food Foodways 15(3–4):213–236. https://doi.org/10.1080/07409710701620136

Maiatsky M (2009) Pirojki à la bolognaise: Le miroir russe de l'université européenne. Multitudes 39(4):100–108. https://doi.org/10.3917/mult.039.0100

Manning JA (2009) Constantly containing. Dissertation, West Virginia University

Maranzano B (2014) Lo sviluppo del fenomeno "street food": il cibo di strada a Palermo ieri e oggi. Dissertation, University of Pisa, Italy

Marino AMF, Giunta R, Salvaggio A, Farruggia E, Giuliano A, Corpina G (2010) Valutazione microbiologica di un prodotto alimentare tipico della tradizione siciliana: l'arancino. Riv Sci Aliment 39(4):17–21. Available https://fosan.it/system/files/Anno_39_4_3.pdf. Accessed 10 April 2020

Matalas AL, Zampelas A, Stavrinos V, Wolinsky I (2001) The Mediterranean diet. Constituents and health promotion. CRC Press, Boca Raton, pp 46–66

Matz SA (1991) The chemistry and technology of cereals as food and feed, 2nd edn. Van Nostrand Reinhold, New York

McHugh MR (2015) Modern palermitan markets and street food in the Ancient Roman World. Conference paper, the Oxford Symposium on Food and Cookery, St. Catherine's College, Oxford University

Messer E, Cohen MJ (2007) Conflict, food insecurity and globalization. Food Cult Soc 10(2):297–315. https://doi.org/10.2752/155280107X211458

Ministero delle politiche agricole alimentari e forestali (2014) Quattordicesima revisione dell'elenco dei prodotti agroalimentari tradizionali. Ministero delle politiche agricole alimentari e forestali, Rome. Available https://www.politicheagricole.it/flex/cm/pages/ServeBLOB.php/L/IT/IDPagina/3276. Accessed 7 April 2020

Mir-Bel J, Oria R, Salvador ML (2013) Reduction in hydroxymethylfurfural content in 'churros', a Spanish fried dough, by vacuum frying. Int J Food Sci Technol 48(10):2042–2049. https://doi.org/10.1111/ijfs.12182

Morton PE (2014) Tortillas: a cultural history. University of New Mexico Press, Albuquerque

Myfitnesspal (2020) Sfincione—Sicilian Pizza. www.myfitnesspal.com. Available https://www.myfitnesspal.com/it/food/calories/sicilian-pizza-13880320. Accessed 8 April 2020

Oliveras-López MJ, Molina JJM, Mir MV, Rey EF, Martín F, de la Serrana HLG (2013) Extra virgin olive oil (EVOO) consumption and antioxidant status in healthy institutionalized elderly humans. Arch Gerontol Geriatr 57(2):234–242. https://doi.org/10.1016/j.archger.2013.04.002

Palladino M (2018) (Im) mobility and Mediterranean migrations: journeys 'between the pleasures of wealth and the desires of the poor.' J North Afr Stud 23(1–2):71–89. https://doi.org/10.1080/13629387.2018.1400241

Parente G (2007) Cibo veloce e cibo di strada. Le tradizioni artigianali del fast-food in Italia alla prova della globalizzazione. Storicamente 3(2). https://doi.org/10.1473/stor389

Parisi S (2019) Analysis of major phenolic compounds in foods and their health effects. J AOAC Int 102(5):1354–1355. https://doi.org/10.5740/jaoacint.19-0127

Parisi S (2020) Characterization of major phenolic compounds in selected foods by the technological and health promotion viewpoints. J AOAC Int, in Press. https://doi.org/10.1093/jaoacint/qsaa011

Parisi S, Delia S, Laganà P (2004) Il calcolo della data di scadenza degli alimenti: la funzione Shelf Life e la propagazione degli errori sperimentali. Ind Aliment 43(438):735–749

Parisi S, Luo W (2018) Chemistry of Maillard reactions in processed foods. Springer, Springer International Publishing, Cham

Parisi S, Ameen SM, Montalto S, Santangelo A (2019) Maillard reaction in foods: mitigation strategies and positive properties. Springer International Publishing, Cham

Pasqualone A, Delcuratolo D, Gomes T (2011) Focaccia Italian flat fatty bread. In: Preedy VR, Watson RR, Patel VB (eds) Flour and breads and their fortification in health and disease prevention. Academic Press, Cambridge. https://doi.org/10.1016/B978-0-12-380886-8.10005-4

Patel K, Guenther D, Wiebe K, Seburn RA (2014) Promoting food security and livelihoods for urban poor through the informal sector: a case study of street food vendors in Madurai, Tamil Nadu, India. Food Sec 6(6):861–878. https://doi.org/10.1007/s12571-014-0391-z

Pignataro L (2018) La pizza: Una storia contemporanea. Hoepli Editore, Milan

Pilcher JM (2017) Planet taco: a global history of Mexican food. Oxford University Press, Oxford

Pitte JR (1997) Nascita e diffusione dei ristoranti. In: Flandrin JL, Montanari M (eds) Storiadell'alimentazione, Laterza, Roma and Bari

Privitera D (2015) Street food as form of expression and socio-cultural differentiation. In: Proceedings of the 12th PASCAL International Observatory Conference, Catania

Pshenichnikova TA, Osipova SV, Permyakova MD, Mitrofanova TN, Trufanov VA, Lohwasser U, Röder M, Börner A (2008) Mapping of quantitative trait loci (QTL) associated with activity of disulfide reductase and lipoxygenase in grain of bread wheat *Triticum aestivum* L. Genet 44(5):654–662. https://doi.org/10.1134/s1022795408050098

Rajendran A (2004) Adsorption and chromatography under supercritical conditions. Dissertation, Swiss Federal Institute of Technology, Zurich

Salloum H, Salloum M, Elias LS (2013) Sweet delights from a thousand and one nights: the story of traditional Arab sweets. I.B. Tauris & Co., Ltd., London and New York

Shackman G, Yu C, Edmunds LS, Clarke L, Sekhobo JP (2015) Relation between annual trends in food pantry use and long-term unemployment in New York state, 2002–2012. Am J Pub Health 105(3):e63–e65. https://doi.org/10.2105/AJPH.2014.302382

Sheen B (2010) Foods of Egypt. Greenhaven Publishing LLC, New York

Silverman C (2012) Education, Agency, and Power among Macedonian Muslim Romani Women in New York City. Signs J Women Cult Soc 38(1):30–36. https://doi.org/10.1086/665803

Simopoulos AP, Bhat RV (2000) Street foods. Karger AG, Basel

Singla RK, Dubey AK, Ameen SM, Montalto S, Parisi S (2018) Analytical methods for the assessment of Maillard reactions in foods. Springer International Publishing, Cham

Steven QA (2018) Fast food, street food: western fast food's influence on fast service food in China. Dissertation, Duke University, Durham

Steyn NP, Labadarios D (2011) Street foods and fast foods: How much do South Africans of different ethnic groups consume? Ethnic Dis 21(4):462–466

Steyn NP, Mchiza Z, Hill J, Davids YD, Venter I, Hinrichsen E, Opperman M, Rumbelow J, Jacobs P (2013) Nutritional contribution of street foods to the diet of people in developing countries: a systematic review. Pub Health Nutr 17(6):1363–1374. https://doi.org/10.1017/S136898001300 1158

Stoica A, Barascu E, Hossu AM (2013) Improving the quality of bread made from "short" gluten flours using a fungal pentosanase and L-cysteine combination. Rev Chim 64(9):951–954. Available https://www.revistadechimie.ro/pdf/STOICA%20ALEX.pdf%209%2013.pdf. Accessed 10 April 2020

Tinker I (1999) Street foods into the 21st century. Agric Human Val 16:327-333

Toktassynova Z, Akbaba A (2017) Content analysis of on-line booking platform reviews over a restaurant: a case of pizza locale in Izmir. Avrasya Sosyal ve Ekonomi Araştırmaları Dergisi 5(5):242–249

Vanschaik B, Tuttle JL (2014) Mobile food trucks: California EHS-Net study on risk factors and inspection challenges. J Environ Health 76(8):36–37. Gale Academic OneFile. Available https://go.gale.com/ps/anonymous?id=GALE%7CA365689844&sid=googleScholar&v=2.1&it=r&linkaccess=abs&issn=00220892&p=AONE&sw=w. Accessed 7 April 2020

Webb RE, Hyatt SA (1988) Haitian street foods and their nutritional contribution to dietary intake. Ecol Food Nutr 21(3):199–209. https://doi.org/10.1080/03670244.1988.9991033

Wilkins J, Hill S (2009) Food in the ancient world. Blackwell Publishing Ltd, Maiden, Oxford, and Carlton

Willett WC, Sacks F, Trichopoulou A, Drescher G, Ferro-Luzzi A, Helsing E, Trichopoulos D (1995) Mediterranean diet pyramid: a cultural model for healthy eating. Am J Clin Nutr 61(6):1402S-1406S. https://doi.org/10.1093/ajcn/61.6.1402S

Chapter 4
Palermo's Street Foods. The Authentic *Pani câ Meusa*

Abstract Street food is not a synonym of Mediterranean-Diet-related foods. An internationally recognised definition of street foods clearly identifies these products as ready-to-eat foods and beverages which are prepared and sold by different food business operators without an immobile location. Related features—including cheapness and nutritional contents—do not seem applicable to the Mediterranean Diet. Certain well-known Sicilian specialities, such as *arancina*, *arancino*, or *sfincionello*, are identified as street foods without connections between their presumptive safety/health and the current status of food produced and sold 'on the road'. On the contrary, these foods are recognised as excellent 'one-food' solutions being able to supply the needed energy and many nutritional factors at the same time. In this ambit, the study of street foods in the world could show some surprise, especially with reference to particular social and geographical/historical areas. This Chapter concerns a specific street food which can be found only in Palermo, Sicily, although some similar recipe may be found in Southern Italy. Because of the important influence of many civilisations in Sicily and the pre-existing Hebraic traditions, the study of a peculiar Palermo's sandwich—the *pani câ meusa* street food—is highly recommended by different viewpoints including history, possible 'authenticity' features, and identification of raw materials, preparation procedures, concomitant alternative recipes, and nutritional facts.

Keywords Beef offal · Beef spleen · Frying · Hebraic cuisine · *Pani câ meusa* · Sicily · Street food

Abbreviations

AFSUN	African Food Security Urban Network
CNC	Consiglio Nazionale dei Chimici
FAO	Food Agriculture Organization of the United Nations
FBO	Food business operator

MD   Mediterranean Diet
SF   Street food
TSG   Traditional speciality guaranteed

4.1 Certain Street Foods Are Found Only in One Place…

As mentioned in Chap. 1, 'street food' (SF) is not a synonym of the Mediterranean-Diet (MD)-related foods. An internationally recognised definition of SF clearly identifies these products as ready-to-eat foods and beverages which are prepared and sold by different food business operators (FBO) without an immobile location. SF can be found only in the streets and in similar places (including also small food trucks) (Anenberg and Kung 2015; FAO 1997; Maranzano 2014; Pitte 1997; Vanschaik and Tuttle 2014).

On the other hand, MD is a synonym of tradition, cultural heritage, food safety, healthy dietary guidelines. These attributions depend on the recognised advantages of MD-foods (low calories; low carbohydrates, sugars, and lipids), and the presence of antioxidants including polyphenols (Haddad et al. 2020a, b; Parisi 2019, 2020).

On these bases, it could be inferred that SFs are not 'healthy' or safe'. Actually, this interpretation is not correct. As explained in Chaps. 2 and 3, certain well-known Sicilian specialities as *arancina*, *arancino*, or *sfincionello*, are identified as SF. At the same time, these products are real 'ready-to-eat' and 'takeaway' foods without connections between their presumptive safety/health and the current status of food produced and sold 'on the road'. On the contrary, these foods are recognised as excellent 'one-food' solutions being able to supply the needed energy and many nutritional factors at the same time. It should be admitted that the 'health danger' is related to the amount of purchased and assumed SF per single person, instead of its composition. Naturally, hygienic conditions of SF vendors are another matter, and this aspect should be considered apart.

It has also been considered that SFs are correlated with social aggregation and urbanisation. The following conditions: economic convenience, lack of immobile locations, 'poor' food features, typical folk elements, and a certain trend towards poor hygienic conditions; are or should be taken into account (Adjrah et al. 2013; Chukuezi 2010; Crush et al. 2011; Fellows and Hilmi 2012; Hawkes et al. 2017; Sujatha et al. 1997). Above all, SFs are preferred in many cities because of their amount of bioavailable proteins and energy intake, and also because of their economic convenience (cheapness). The higher the price, the higher the inherent value of the food.

In this ambit, the study of SF in the world could show some surprise, especially with reference to particular social and geographical/historical areas. This book is dedicated to the study of a peculiar type of folk cuisine found in the Sicilian largest city, Palermo. In particular, this Chapter concerns a specific SF which can be found only in Palermo, and its history is linked to this area only, although some similar recipe may be found in another city of Southern Italy, Naples.[1] Because of the important influence of many of these civilisations in Sicily and in Naples (Maranzano 2014; Anonymous 2020a), these situations may be expected. Anyway, it is possible to find typical SF of the Palermo area, but one or two of these products are explicitly found in Palermo only: the *pani câ meusa* SF.

4.2 What is *Pani câ Meusa*?

The *pani câ meusa* SF is a peculiar feature found only in Palermo, Sicily, and its presence is not recognised on other Sicilian areas and worldwide. Such a similar presence may be questionable. In general, SFs with a solid tradition and historical heritage are well-known in many different areas at the same time (Chaps. 2 and 3). A little example is the coexistence of a Sicilian sweet product–*'mpanatigghi'*–and its corresponding (and probably original) version in Spain, *'empanadillas'*. A similar product of Puerto Rico (*'empanadillas'*) can be reported (Bacarella 2003; Castiglione 2011; Constable 2013; Cuesta and Ainciburu 2015; Fernández-Coll 1985; García et al. 2009; Roden 2016; Ruggiero and Scrofani 2009). It is interesting to note that all of these products are derivations of the Muslim dominations in both the Hispanic peninsula and Sicily, and subsequently of the Spanish Empire (in Puerto Rico). Anyway, one single recipe can be easily found with more or less notable differences in many places. On the other hand, the *pani câ meusa* appears very different. Certainly, this food is classifiable only as SF, while normal catering services or restaurants usually do not serve it.

What is exactly *pani câ meusa*? Substantially, it is a peculiar sandwich filled with different ingredients including cow spleen. The meaning of the Sicilian name is now evident: *pani* is for 'bread' (the original roll, here also named *vastedda*) and *meusa* is for 'beef spleen'. Literally, the Sicilian name is 'bread sandwich with beef spleen'. The above-mentioned description is too simplistic, even with relation to the simple name... We have found the following definitions for this SF (Buscemi et al. 2015; Marcato 2017; Petrini 2003; Varisco 2013):

- *Panino con milza* (Italian language),
- Cow spleen sandwich (English language),
- Beef spleen sandwich (English language),
- *Pani ca' meusa* (Sicilian dialect),
- *Pane ca meusa* (Italian language and Sicilian dialect),
- *Pani câ meusa* (Sicilian dialect).

With concern to this book, we will use the last name. Anyway, all of these descriptions concern always the same Palermo's food.

Historically, this SF speciality is found only in the Sicilian capital city, Palermo, on the western side, and with two versions. This and other similar foods may be

---

[1]This recipe is named '*Milza alla Napoletana*' or 'Naples spleen'. However, this is not a sandwich.

defined as a cultural heritage of the Sicilian cuisine (Barbagli and Barzini 2010; CNC 2014a, b, c, d, e). However, the *pani câ meusa* should not be considered as derived from Muslim cuisine. More probably, it should have been created during the Aragonese period in Sicily and before the discovery of the New World, between 1412 and 1492 A.D. (Woodacre 2013). This information is deduced from a project concerning the application for registration of a 'traditional speciality guaranteed' (TSG), according to the Title III of the Regulation (EU) No. 1151/2012 of the European Parliament and of the Council of 21 November 2012 on quality schemes for agricultural products and foodstuffs (European Parliament and Council 2012). This project has been proposed in Italy with relation to '*pane ca meusa*' (CNC 2014d). This proposal followed the creation of a '*City Brand Panormvs*' for the protection of Palermo's products (Maranzano 2014). The TSG proposal aims at the creation of the '*Panormvs—street food*' brand specifically linked to the TGS product(s) and the realisation of an international SF festival in Palermo (CNC 2014a, b). At present, there is a 'Panormus Street Food' SF mobile caterer company currently working in the United Kingdom (Andolina 2019). On the other side, there is not available information concerning the *pani câ meusa* TSG: there are not Italian applications in this ambit as TSG on 24 March 2020 (European Commission 2019, 2020).

According to this reference, one of the basic ingredients for this SF is lard or *saimi*, generally extracted after pork slaughter. This fat matter is named similarly to Spanish *saim*. It may be assumed that the extraction and the purification of this product are concomitant with the creation of a sandwich-like food composed of unleavened bread containing beef offal. It has to be clarified that the Hebraic tradition forces the slaughterer to keep only beef offal with the exclusion of liver (no money payment), but Hebrews cannot eat these materials. Consequently, the creation of this sandwich-like food would be the attempt to preserve offal and offer them to non-Hebrew citizens into unleavened bread. After the departure of Hebrews from Palermo, the initial recipe would have been subjected to a few modifications only with the current aspect of *pani câ meusa* (Basile 2015): in detail, bread is now leavened as usual, but the soft inner part of the sandwich is removed (similarly to Hebraic unleavened bread)…

4.3 *Pani câ Meusa*. Appearance, Ingredients, and Preparation

4.3.1 *General Procedures and Ingredients*

As above explained, this SF is similar to a normal sandwich. In detail (Fig. 4.1):

(a) The bread (*vastedda*) is superficially coated with sesame seeds. Similarly to *arancina* SF (Chap. 1), the diameter ranges from 8 to 10 cm, and the same thing is considered with concern to weight (150–200 g).

pani câ meusa schitta   *pani câ meusa maritata*

Fig. 4.1 The traditional *pani câ meusa*. This SF is similar to a normal sandwich. The bread (*vastedda*) is superficially coated with sesame seeds (diameter: 8–10 cm, weight: 150–200 g). The filling contains sautéed beef offal (lung and spleen, with optional beef trachea) One version considers this filling with addition of lemon juice (*pani câ meusa schitta*). Otherwise, the sautéed filling may be composed of a little amount of fresh sheep's *ricotta* or seasoned *caciocavallo* cheese (*pani câ meusa maritata*)

(b) The filling of this sandwich contains only: beef offal (lung and spleen, approximately in the 2:1 ratio, with an optional little amount of beef trachea). The filling has to be prepared with these cooked parts, previously treated in boiling water, and subsequently sliced as thin stripes. The filling has to be sautéed with lard into a special stainless container (this browning or sautéing method should be carried out within one minute). One version considers this filling with the addition of lemon juice. The obtained product will be named *pani câ meusa schitta* (where *schitta* means 'simple'). Otherwise, the sautéed filling may be composed of a little amount of fresh sheep's *ricotta* or seasoned *caciocavallo* cheese (instead of lemon juice). In this ambit, the obtained product will be named *pani câ meusa maritata* (where *maritata* means 'married' or joint with *ricotta* or cheese).

(c) The finished product should show the filling as composed of grey-coloured, soft, and easily chewable offal. This point is critical and dependent on the final browning procedure (time should not exceed 60 s).

As a result, allowed ingredients have to be according to the above-mentioned TSG proposal (CNC 2014d):

(1) *Vastedda* preparation: wheat flour 00, brewer's yeast, water, sea salt, and sesame seeds (CNC 2014e). It has been recognised as an Italian traditional food product or '*prodotto agroalimentare tradizionale italiano*' (PAT) of the Italian Ministry of Agricultural, Food and Forestry Policies (MiPAAF) (Ministero delle politiche agricole alimentari e forestali 2014).

(2) Filling preparation (bases): beef lung, beef spleen, beef trachea (if available and desired), sea salt, and purified lard.

(3) Filling types:

(3.1) *Schitta* version: addition of lemon juice,

(3.2) *Maritata* version: sheep's *ricotta* or *caciocavallo* cheese of Godrano (also named Palermo's *caciocavallo* cheese).

4.3.2 Additional Notes

4.3.2.1 Beef Offal

Beef offal has to be maintained in a good state. For this reason, sea salt is strictly required. In addition, the cooking treatment for offal has to be carried out until enzymatic denaturation of meats is obtained (recommended temperature: 80 °C). In these conditions, blood is no longer visible on 15–20 mm-thickness stripes and these materials may be stored in refrigerated conditions (2–4 °C) until four days. Temperature for enzymatic denaturation is 80 °C.

4.3.2.2 Filling Sautéed in Lard

This critical step is carried out by SF vendors into a stainless container with a diameter of 50 cm and height approximately 25 cm. This container is inclined if compared with the position of the used burner. This factor is important because the browning process has to maintain beef offal in the upper side of the filled container, while melted lard has to remain on the bottom. After one minute only, separated meat stripes appear joint and can be placed into the opened *vastedda*. In this way, the exceeding fat (lard) will easily get out of the sandwich and return into the original container when the sandwich will be turned (inclined) by hand before serving.

4.3.2.3 *Schitta* or *Maritata* Versions

As above explained, two *pani câ meusa* versions are available. It has to be clarified that the possible addition of *ricotta* or *caciocavallo* cheese has to be performed on the upper surface of inner *vastedda* slice. In this way (*maritata* version), the filling is deposed exactly on a cheese or *ricotta* 'bed' directly onto bread surface. Alternatively (*schitta* version), only lemon juice is added to bread. Anyway, the prepared sandwich has to be 'turned' (inclined) and pressed by hand with the aim of getting out exceeding melted lard from the sandwich. The final serving requires abundant greaseproof paper sheets.

4.4 Optional Ingredients

The above-mentioned TSG proposal shows approximately all key factors of *pani câ meusa* SF. Other considerations have to be mentioned before concluding this Chapter with nutritional data.

Some reference may suggest the use of lemon juice and pepper together. In this ambit, the *pani câ meusa* version would not be called *schitta*: differently from the TSG proposal, the simple version should contain filling only (Cardella 2009). At the same time, the SF vendor is accustomed to add sea salt into the sautéed filling just before composing the final sandwich (Di Giorgi 2017). This addition is mentioned in the TSG proposal only when speaking of initial meat cooking.

4.5 Nutritional Profile of *Pani câ Meusa* SF

The above-mentioned description of *pani câ meusa* SF appears detailed enough when speaking of amounts concerning raw materials and ingredients, differently from *arancina* (Chap. 1) or *sfincionello* (Chap. 2). However, there are not many available data concerning the nutritional profile of this food.

We can use only two available references (Anonymous 2020b; Basile 2015), taking into account the following facts:

(1) Nutritional data are only an indication concerning a product which has not been considered for analytical purposes until recent times.
(2) The only available data are referred to a single 250 g-sandwich. This note has to be taken into account because of the important amount of water (125 g).
(3) There are not scientific papers concerning related durability in our knowledge.

Taking into account the above-mentioned premise, the following nutrition facts per 250 g should be assumed for *pani câ meusa* SF:

- Energy: 697 kcal per 250 g,
- Lipids: 41 g,
- Saturated fat: 15 g,
- Carbohydrates: 57 g,
- Sugars: unavailable data,
- Fibres: 4 g,
- Protein: 27 g,
- Sodium: 662 mg,
- Cholesterol: 34 mg.

In addition, water is considered approximately equal to 125 g per 250 g of this SF. Taking into account that the *vastedda* bread is approximately 150–200 g, it can be assumed that the remaining matters range from 50 to 100 g.

On these bases, some of these data appear questionable (Table 4.1):

Table 4.1 Nutrition facts for *pani câ meusa* SF based on two web references (Anonymous 2020b; Basile 2015)

Nutritional profile	Nutritional profile per 250 g	Nutrition facts per 100 g
Energy (kcal)	697	279
Lipids	41	16.4
Saturated fat	15	6
Carbohydrates	57	22.8
Sugars	–	–
Fibres	4	1.6
Protein	27	10.8
Sodium	662	265
Cholesterol	34	14
Re-calculated energy (kcal)[a]	705	282

Interestingly, some data appear questionable taking into account a 250 g weight for a single sandwich. The Table shows two columns series: the original data per 250 g, and nutrition facts per 100 g, with the re-calculated energy intake (as kcal). Some differences appear to show that original data are not coherent enough, while the sum of the three main nutrients (lipids, carbohydrates, and protein) is exactly the dry matter (250–125 g of water = 125 g)

[a]Based on available data concerning lipids, carbohydrates, and protein

(a) The calculated energy intake (as kcal) on the basis of listed data (lipids, carbohydrates, and protein) is 705 instead of 697 kcal

(b) The interpolated data per 100 g show another difference (282 instead of theoretical 278 kcal).

On a general level, the main amount of all expressed nutrients concerns carbohydrates; however, above-expressed doubts should be taken into account. Consequently, the energy intake of lipids (16.4% against 22.8% for carbohydrates) should be re-evaluated. After all, energy from lipids is calculated as 'grams × 9' kcal.

A final note concerning these data is related to the declared sum of carbohydrates, lipids, and protein. This sum is exactly 125 g, or the total dry matter (Table 4.1, first column). In this situation, should carbohydrates and protein be correct, the amount of lipids would be approximately 40.1 instead of 41 g. Should carbohydrates and lipids be correct, the amount of protein would be 25 instead of 27 g. Finally, should lipids and protein be correct, the amount of carbohydrates would be 55 instead of 57 g. All the above shown differences may be explainable in terms of simple analytical errors. On the other hand, the sum of the main nutritional amounts in foods rarely corresponds to the dry matter. Consequently, more research should be needed with relation to *pani câ meusa* SF. On the basis of available data, there are two differences between this and other Palermo's SF (Chaps. 1 and 2) concerning the nutritional amount of proteins and lipids. This SF contains a notable amount of animal proteins and fat, and this fact (with the absence of vegetable oil!) has to be considered as the best nutritional expression of *pani câ meusa* SF...

References

Adjrah Y, Soncy K, Anani K, Blewussi K, Karou DS, Ameyapoh Y, de Souza C, Gbeassor M (2013) Socio-economic profile of street food vendors and microbiological quality of ready-to-eat salads in Lomé. Int Food Res J 20(1):65–70

Andolina C (2019) The Panormus Street food: la rosticceria palermitana conquista i mercati inglesi. www.orogastronomico.it. Available https://www.orogastronomico.it/news-ed-eventi/the-panormus-street-food-la-rosticceria-palermitana-in-inghilterra/. Accessed 7 April 2020

Anenberg E, Kung E (2015) Information technology and product variety in the city: the case of food trucks. J Urban Econ 90:60–78. https://doi.org/10.1016/j.jue.2015.09.006

Anonymous (2020a) Come cucinare la milza. www.galbani.it. Available https://www.galbani.it/abc ucina/come-fare/come-cucinare-la-carne/come-cucinare-la-milza. Accessed 8 April 2020

Anonymous (2020b) Cibo Da Strada - Palermo - Panino Con La Milza. www.myfitness pal.com. Available https://www.myfitnesspal.com/food/calories/cibo-da-strada-palermo-panino-con-la-milza-13844566. Accessed 8 April 2020

Bacarella A (ed) (2003) Agroalimentare e flussi turistici in Sicilia. University of Palermo, Palermo

Barbagli A, Barzini S (2010) Pane, pizze e focacce. Giunti Editore, Florence

Basile G (2015) Piaceri e misteri dello street food palermitano. Storia, aneddoti e sapori del cibo di strada più buono d'Europa. Dario Flaccovio Editore, Palermo. ISBN 978-88-579-0461-0

Buscemi S, Rosafio G, Vasto S, Massenti FM, Grosso G, Galvano F, Rini N, Barile AM, Maniaci V, Cosentino L, Verga S (2015) Validation of a food frequency questionnaire for use in Italian adults living in Sicily. Int J Food Sci Nutr 66(4):426–438. https://doi.org/10.3109/09637486.2015.102 5718

Cardella G (2009) Pane con la milza. www.ricettedisicilia.net. Available https://www.ricettedisic ilia.net/piatti-unici/pane-con-la-milza-"pani-c'a-mieusa"/. Accessed 7 April 2020

Castiglione M (2011) Tradizione, identità, tipicità nella cultura alimentare siciliana. Lo sguardo dell'ALS. Centro Studi Filologici e Linguistici Siciliani, Palermo

Chukuezi CO (2010) Food safety and hyienic practices of street food vendors in Owerri, Nigeria. Stud Sociol Sci 1(1):50–57. https://doi.org/10.3968/j.sss.1923018420100101.005

CNC (2014a) Accordo tra Comune di Palermo e Consiglio Nazionale dei Chimici (CNC) per la partecipazione alla realizzazione di un sistema di salvaguardia e garanzia della tradizione gastronomica palermitana', Prot. 646/14/cnc/fta. Consiglio Nazionale dei Chimici (CNC), Rome. Available https://www.chimicifisici.it/wp-content/uploads/2018/10/20131210_accordo_firmato_dal_Presidente_del_CNC.pdf. Accessed 7 April 2020

CNC (2014b) DOMANDA DI REGISTRAZIONE - Art. 8 - Regolamento (UE) n. 1151/2012 del Parlamento Europeo e del Consiglio del 21 novembre 2012 sui regimi di qualità dei prodotti agricoli e alimentari - "ARANCINA". Annex to the document 'Accordo tra Comune di Palermo e Consiglio Nazionale dei Chimici (CNC) per la partecipazione alla realizzazione di un sistema di salvaguardia e garanzia della tradizione gastronomica palermitana', Prot. 646/14/cnc/fta. Consiglio Nazionale dei Chimici (CNC), Rome. Available https://www.chimicifisici.it/wp-con tent/uploads/2018/10/ARANCINA_CNC__STG_2014.pdf. Accessed 8 April 2020

CNC (2014c) DOMANDA DI REGISTRAZIONE - Art. 8 - Regolamento (UE) n. 1151/2012 del Parlamento Europeo e del Consiglio del 21 novembre 2012 sui regimi di qualità dei prodotti agricoli e alimentari - "SFINCIONELLO". Annex to the document 'Accordo tra Comune di Palermo e Consiglio Nazionale dei Chimici (CNC) per la partecipazione alla realizzazione di un sistema di salvaguardia e garanzia della tradizione gastronomica palermitana', Prot. 646/14/cnc/fta. Consiglio Nazionale dei Chimici (CNC), Rome. Available https://www.chimicifisici.it/wp-con tent/uploads/2018/10/SFINCIONELLO_CNC_STG_2014.pdf. Accessed 7 April 2020

CNC (2014d) DOMANDA DI REGISTRAZIONE - Art. 8 - Regolamento (UE) n. 1151/2012 del Parlamento Europeo e del Consiglio del 21 novembre 2012 sui regimi di qualità dei prodotti agricoli e alimentari - "PANE CA MEUSA". Annex to the document 'Accordo tra Comune di Palermo e Consiglio Nazionale dei Chimici (CNC) per la partecipazione alla realizzazione di un sistema di salvaguardia e garanzia della tradizione gastronomica palermitana', Prot. 646/14/cnc/fta.

Consiglio Nazionale dei Chimici (CNC), Rome. Available https://www.chimicifisici.it/wp-con tent/uploads/2018/10/PANE_CA_MEUSA_rev.3dic2014.pdf. Accessed 8 April 2020

CNC (2014e) DOMANDA DI REGISTRAZIONE DI UNA STG - Art. 8 - Regolamento (UE) n. 1151/2012 del Parlamento Europeo e del Consiglio del 21 novembre 2012 sui regimi di qualità dei prodotti agricoli e alimentary - "PANE E PANELLE". Annex to the document 'Accordo tra Comune di Palermo e Consiglio Nazionale dei Chimici (CNC) per la partecipazione alla realiz-zazione di un sistema di salvaguardia e garanzia della tradizione gastronomica palermitana', Prot. 646/14/cnc/fta. Consiglio Nazionale dei Chimici (CNC), Rome. Available https://www.chimic ifisici.it/wp-content/uploads/2018/10/PANE_E_PANELLE__CNC_STG_2014.pdf. Accessed 7 April 2020

Constable OR (2013) Food and meaning: Christian understandings of Muslim food and food ways in Spain, 1250–1550. Viator 44(3):199–235. https://doi.org/10.1484/J.VIATOR.1.103484

Crush J, Frayne B, McLachlan M (2011) Rapid urbanization and the nutrition transition in Southern Africa. Urb Food Secur Netw Series 7:1–49. Queen's University and African Food Security Urban Network (AFSUN), Kingston and Cape Town. Available https://fsnnetwork.org/sites/def ault/files/rapid_urbanization_and_the_nutrition.pdf. Accessed 7 April 2020

Cuesta AR, Ainciburu MC (2015) Transfer of Arabic formulaic courtesy expressions used by advanced Arab learners of Spanish. Procedia Soc Behav Sci 173:207–213. https://doi.org/10. 1016/j.sbspro.2015.02.054

Di Giorgi V (2017) Lo Spleen di Palermo. Rappresentazioni e identità intorno al cibo di strada palermitano. Dissertation, Università Ca'Foscari, Venice

European Commission (2019) DOOR database. European Commission, Brussels. Avail-able https://ec.europa.eu/agriculture/quality/door/list.html?recordStart=0&recordPerPage=10& recordEnd=10&sort.mileston. Accessed 7 April 2020

European Commission (2020) eAmbrosia—the EU geographical indications register. Euro-pean Commission, Brussels. Available https://ec.europa.eu/info/food-farming-fisheries/food-saf ety-and-quality/certification/quality-labels/geographical-indications-register/. Accessed 7 April 2020

European Parliament and Council (2012) Regulation (EU) No 1151/2012 of the European Parlia-ment and of the Council of 21 November 2012 on quality schemes for agricultural products and foodstuffs. OJ Eur Union L 343:1–29

FAO (1997) Street foods. FAO Food and Nutrition Paper N. 63. Report of an FAO Technical Meeting on Street Foods, Calcutta, India, 6–9 November 1995. Food and Agriculture Organization of the United Nations (FAO). Available https://www.fao.org/3/W4128T/W4128T00.htm/. Accessed 7 April 2020

Fellows P, Hilmi M (2012) Selling street and snack foods. FAO Diversification booklet number 18. Rural Infrastructure and Agro-Industries Division, Food and Agriculture Organization of the United Nations (FAO), Rome. Available https://www.fao.org/docrep/015/i2474e/i2474e00.pdf. Accessed 7 April 2020

Fernández-Coll F (1985) Microbiological quality of some Puerto Rican fast foods. I. Processor level; Frozen or refrigerated. J Agric Univ Puerto Rico 69(1):81–89

García MM, Seiquer I, Delgado-Andrade C, Galdó G, Navarro MP (2009) Intake of Maillard reaction products reduces iron bioavailability in male adolescents. Mol Nutr Food Res 53(12):1551–1560. https://doi.org/10.1002/mnfr.200800330

Haddad MA, Abu-Romman S, Obeidat M, Iommi C, El-Qudah J, Al-Bakheti A, Awaisheh S, Jaradat DMM (2020a) Phenolics in Mediterranean and Middle East important fruits. J AOAC Int, in press

Haddad MA, Obeidat M, Al-Abbadi A, Shatnawi MA, Al-Shadaideh A, Al-Mazra'awi MI , Iommi C, Dmour H, Al-Khazaleh JM (2020b) Herbs and medicinal plants in Jordan. J AOAC Int, in press

Hawkes C, Harris J, Gillespie S (2017) Urbanization and the nutrition transition. Glob Food Pol Rep 4:34–41. https://doi.org/10.2499/9780896292529_04

Maranzano B (2014) Lo sviluppo del fenomeno "street food": il cibo di strada a Palermo ieri e oggi. Dissertation, University of Pisa, Italy

Marcato G (2017) Dialetto – Uno Nessuno Centomila. Coop. Libraria Editrice Università di Padova (CLEUP), Padova

Ministero delle politiche agricole alimentari e forestali (2014) Quattordicesima revisione dell'elenco dei prodotti agroalimentari tradizionali. Ministero delle politiche agricole alimentari e forestali, Rome. Available https://www.politicheagricole.it/flex/cm/pages/ServeBLOB.php/L/IT/IDP agina/3276. Accessed 7 April 2020

Parisi S (2019) Analysis of major phenolic compounds in foods and their health effects. J AOAC Int 102(5):1354–1355. https://doi.org/10.5740/jaoacint.19-0127

Parisi S (2020) Characterization of major phenolic compounds in selected foods by the technological and health promotion viewpoints. J AOAC Int, in press. https://doi.org/10.1093/jaoacint/qsaa011

Petrini C (2003) Slow food: the case for taste. Columbia University Press, New York

Pitte JR (1997) Nascita e diffusione dei ristoranti. In: Flandrin JL, Montanari M (eds) Storiadell'alimentazione, Laterza, Roma and Bari

Roden C (2016) The food of Spain. Penguin, London, UK

Ruggiero V, Scrofani L (eds) (2009) Turismo nautico e distretti turistici siciliani. Franco Angeli, Rome

Sujatha T, Shatrugna V, Rao GN, Reddy GCK, Padmavathi KS, Vidyasagar P (1997) Street food: an important source of energy for the urban worker. Food Nutr Bull 18(4):1–5. https://doi.org/10.1177/156482659701800401

Vanschaik B, Tuttle JL (2014) Mobile food trucks: California EHS-Net study on risk factors and inspection challenges. J Environ Health 76(8):36–37. Gale Academic OneFile. Available https://go.gale.com/ps/anonymous?id=GALE%7CA365689844&sid=googleScholar&v=2.1&it=r&linkaccess=abs&issn=00220892&p=AONE&sw=w. Accessed 7 April 2020.

Varisco S (2013) Le diaspore italiane: il caso dei Sicilian American. Dissertation, Università degli Studi dell'Insubria, Varese

Woodacre E (2013) Blanca, Queen of Sicily and Queen of Navarre: connecting the Pyrenees and the Mediterranean via an Aragonese Alliance. In: Woodacre E (ed) Queenship in the Mediterranean. Palgrave Macmillan, New York, pp 207–227. https://doi.org/10.1057/9781137362834_11

Chapter 5
Palermo's Street Foods. The Authentic *Pane e Panelle*

Abstract 'Street foods' are generally correlated with social aggregation, urbanisation, cheapness, and nutritional contents at the same time. In particular, the following conditions: economic convenience; lack of immobile locations; 'poor' food features; typical folk elements; and a certain trend towards poor hygienic conditions; are or should be taken into account. Above all, street foods are extremely preferred in certain areas because of their amount of bioavailable proteins and energy intake, and also because of their economic convenience. The higher the price, the higher the inherent value of the food. In this ambit, the study of street foods in the world could show some surprise. This book is dedicated to the study of a peculiar type of folk cuisine found in the Sicilian largest city, Palermo. In particular, this Chapter concerns a specific product which can be found only in Palermo, although some similar recipe may be found in Northern Italy (Liguria Region). Because of the important influence of many civilisations in Sicily, these situations may be expected. One or two of these products are explicitly found in Palermo, and one of these foods—the *pane e panelle* sandwich—is discussed in this Chapter by different viewpoints including history, possible 'authenticity' features, the identification of raw materials, preparation procedures, concomitant alternative recipes, and nutritional facts.

Keywords Chickpea flour · Cultural heritage · Frying · Mediterranean diet · Genoese cuisine · *Pane e panelle* · Street food

Abbreviations

CNC	Consiglio Nazionale dei Chimici
FAO	Food Agriculture Organization of the United Nations
FBO	Food business operator
MD	Mediterranean Diet
MiPAAF	Ministero delle politiche agricole alimentari e forestali
SF	Street food
TSG	Traditional speciality guaranteed

5.1 An Introduction to… *Pane e Panelle*

As mentioned in Chap. 1, 'street food' (SF) is not a synonym of the Mediterranean-Diet (MD)-related foods. An internationally recognised definition of SF clearly identifies these products as ready-to-eat foods and beverages which are prepared and sold by different food business operators (FBO) without an immobile location. SF can be found only in the streets and in similar places (including also small food trucks) (Anenberg and Kung 2015; FAO 1997; Maranzano 2014; Pitte 1997; Vanschaik and Tuttle 2014).

On the other hand, MD is a synonym of tradition, cultural heritage, food safety, and healthy dietary guidelines. These attributions depend on the recognised advantages of MD-foods (low calories; low carbohydrates, sugars, and lipids), and the presence of antioxidants including polyphenols (Haddad et al. 2020a, b; Parisi 2019, 2020).

On these bases, it could be inferred that SF is not 'healthy' or safe'. Actually, this interpretation is not correct. As explained in Chaps. 2 and 3, certain well-known Sicilian specialities as *arancina*, *arancino*, or *sfincionello*, are identified as SF. At the same time, these products are real 'ready-to-eat' and 'takeaway' foods without connections between their presumptive safety/health and the current status of food produced and sold 'on the road'. On the contrary, these foods are recognised as excellent 'one-food' solutions being able to supply the needed energy and many nutritional factors at the same time. Hygienic conditions of SF vendors are another matter, and this aspect should be considered apart.

It has also been considered that SF is correlated with social aggregation and urbanisation. The following conditions: economic convenience, lack of immobile locations, 'poor' food features, typical folk elements, and a certain trend towards poor hygienic conditions—are or should be taken into account (Adjrah et al. 2013; Chukuezi 2010; Crush et al. 2011; Fellows and Hilmi 2012; Hawkes et al. 2017; Sujatha et al. 1997). Above all, SF is extremely preferred in certain areas because of their amount of bioavailable proteins and energy intake, and also because of their economic convenience (cheapness). The higher the price, the higher the inherent value of the food.

In this ambit, the study of SF in the world could show some surprise, especially with reference to determined social and geographical/historical areas. This book is dedicated to the study of a peculiar type of folk cuisine found in the Sicilian largest city, Palermo. In particular, this Chapter concerns a specific SF which can be found only in Palermo, although some similar recipes may be found in Northern Italy (Liguria Region). Because of the important influence of many civilisations in Sicily, these situations may be expected. Anyway, it is possible to find typical SF of the Palermo area, but one or two of these products are explicitly found in Palermo only: the *pani câ meusa* (discussed in Chap. 4) and the *Pane e Panelle* SF, discussed in this Chapter.

5.2   What Is *Pane e Panelle*? History of a Peculiar Sicilian Food

The *pane e panelle* SF is a peculiar feature found only in Palermo, Sicily, and its presence is not recognised in other Sicilian areas and worldwide. Similarly to the *pani câ meusa* SF (Chap. 4), such a similar presence could appear questionable. In general, SF with a solid tradition and historical heritage (Chaps. 2 and 3) are well known in many different areas at the same time. With relation to *pane e panelle*, it is reported that Genoese traditions show the presence of a peculiar chickpea flatbread named *farinata di ceci* (northern Italy), also named *fainé*. Probably, *pane e panelle* may be considered as a cultural contamination by Genoese merchants in Palermo (where the Genoese community had their own road, a church, and also their market) (Abulafia 1978; Carrabino 2017; CNC 2014b; Dauverd 2006; Guigoni 2004; Pisano 2011).

Anyway, one single recipe can be easily found with more or less notable differences in many places. On the other hand, the *pane e panelle* food seems different. Certainly, this product is classifiable as SF, although normal catering services, restaurants, and also some food industry can serve it.

However, it has to be recognised that a similar version of *pane e panelle* exists in Italy, and precisely in the Liguria Region (Scotto 2010; Tavella 2019). With the exception of *farinata di ceci* dishes, the Savona's *panino con fette* (sandwich with slices) is really similar. The Genoese origin of Savona's *fette* (*panissa*) is really similar to Palermo's *panelle*. As a consequence, the use of *panissa* ad slices into sandwiches becomes the Savona's *panino con fette*.

What is exactly *pane e panelle*? Substantially, it is a peculiar sandwich filled with four or more rectangular slices prepared as a single food: the so-called *panella*.

The meaning of the Sicilian name is now evident: *pane* is for 'bread' (the original roll, here also named *vastedda*) and *panella* means a peculiar food which can be eaten alone. The basic ingredients for *panella* are: chickpea flour; fresh parsley; sea salt; and seeds oil (Sect. 5.3.2.1). The appearance of this 'filling' for the final sandwich is a rectangular shape (dimensions: approximately 90 cm × 65 cm, thickness: 3 mm; weight: 15–20 g).

Literally, the Sicilian name is 'bread sandwich with *panella* (slices)'. The above-mentioned description is too simplistic, even with relation to the simple name. Anyway, there is no doubt that the specific *pane e panelle* SF cannot be found with other names. Consequently, we will use this name.

Historically, this SF speciality is found only in the Sicilian capital city, Palermo, on the western side. This and other similar foods may be defined as a cultural heritage of the Sicilian cuisine (Barbagli and Barzini 2010). In particular, *panella* has been recognised as one of the Italian traditional food products or '*prodotti agroalimentari tradizionali italiani*' (PAT) of the Italian Ministry of Agricultural, Food and Forestry Policies (MiPAAF) (Ministero delle politiche agricole alimentari e forestali 2014). In this ambit, and similarly for other recognised SF of the Palermo traditional cuisine

(Chaps. 2, 3 and 4), a project concerning the application for registration of a 'traditional speciality guaranteed' (TSG), according to the Title III of the Regulation (EU) No. 1151/2012 of the European Parliament and of the Council of 21 November 2012 on quality schemes for agricultural products and foodstuffs (European Parliament and Council 2012) has been proposed in Italy in 2014, with relation to *pane e panelle* (CNC 2014a, b). This proposal followed the creation of a '*City Brand Panormvs*' for the protection of Palermo's products (Maranzano 2014). The TSG proposal aims at the creation of the '*Panormvs—street food*' brand specifically linked to the TGS product(s) and the realisation of an international SF festival in Palermo (CNC 2014a). At present, there is a 'Panormus Street Food' SF mobile caterer company currently working in the United Kingdom (Andolina 2019). On the other side, there is no available information concerning the *pane e panelle* TSG: there are no Italian applications in this ambit as TSG on 24 March 2020 (European Commission 2019, 2020).

Apparently, it could be admitted that *pane e panelle* are similar to *pani câ meusa* (Chap. 4) because both SF are substantially sandwiches. Actually, only the external bread 'container' (the so-called *vastedda* or *guastedda*) is equal in both products, while the filling—*panella* slices—is completely different.

In detail, *panella* slices are obtained from cooked chickpea flour; the subsequent step (frying) has to be performed with the use of seeds oil. As a consequence:

(a) *Pani câ meusa* is based on meat protein and fat (lard), while
(b) *Pane e panelle* is based on chickpea flour and vegetable oil.

This difference (animal vs vegetable ingredients) is the main factor in this ambit.

The historical reason for chickpea flour in Sicily is probably linked with the Arab period of the Sicily Emirate. However, a well-known famine in 1030 A.D. (after the Muslim civilisation) is traditionally linked with the first and documented use of chickpea flour: chickpeas, *Cicer arietinium*, are widely diffused in the whole Mediterranean basin (Duby 1976). The creation of *panella* slices should be linked with the subsequent dominations in Sicily, and probably because of the presence of a well-documented Genoese community in Palermo. It is well known that Genoese traditional cuisine has developed some chickpea recipe such as *farinata di ceci* (the traditional chickpea flatbread in Liguria, northern Italy), also named *fainé*. Probably, this circumstance has originated a local version in Palermo (Abulafia 1978; Carrabino 2017; Dauverd 2006; Guigoni 2004; Pisano 2011). Anyway, the real reason of *panella* foods—and consequent *pane e panelle* SF—should be linked with poor conditions of the common people in the Palermo area of Middle Age and the opportunity of consuming fried fish and seafood 'scraps'. Maybe, for this reason, a common name for ancient *panella* was *piscipanelli* (fish slices). Traditionally, the maximum consumption of *pane e panelle* (and other Palermo's SF such as *arancina*) is temporally located in early December each year, and particularly at the anniversary of St. Lucy (13th December). The traditional explanation is that the consumption of wheat-derived foods is avoided on this date.

5.3 *Pane e Panelle.* **Appearance, Ingredients, and Preparation**

5.3.1 *General Features and Ingredients*

As above explained, this SF is similar to a normal sandwich (CNC 2014b). In detail (Fig. 5.1):

(a) The bread (*vastedda*) is superficially coated with sesame seeds (Sect. 4.3.1). Similarly to *arancina* SF (Chap. 1), the diameter ranges from 8 to 10 cm, and the same thing is considered with concern to weight (200 g).
(b) The filling of this sandwich contains only: *panella* fried slices, although an alternative version considers also the mixed use of *panella* and fried croquettes (named *crocchè*).
(c) Optional ingredients may be ground black pepper and fennel seeds
(d) The finished product should show the filling as composed of yellow-coloured, soft and detached fried *panella* slices with the following chromatic features: Cyan, 0%; Magenta, 14%; Yellow, 10%; Black, 0%. This point is critical and dependent on the frying procedure (browning or sautéing method—time should not exceed 60 s).

As a result, allowed ingredients have to be—according to the above-mentioned TSG proposal (CNC 2014b):

(1) *Vastedda* preparation: wheat flour 00, brewer's yeast, water, sea salt, and sesame seeds.

The *Pane e panelle* speciality   *Panella* slices

Vastedda

Fig. 5.1 The *pane e panelle* speciality. This SF is similar to a normal sandwich, where the bread (*vastedda*) is superficially coated with sesame seeds (diameter: 8–10 cm, weight is approximately 200 g). The filling contains only fried *panella* slices, although an alternative version considers also the mixed use of *panella* and fried croquettes. Optional ingredients: ground black pepper and fennel seeds

(2) Filling preparation: fried *panella* slices (with an optional possibility: the addition of potato croquettes, called *crocchè*). Ingredients for *panella* slices are: chickpea flour; fresh parsley; sea salt; seeds oil.

5.3.2 Procedures

5.3.2.1 *Panella* Production

In accordance with the above-mentioned TSG proposal, a mixture has to be prepared by dissolving 1000 g of chickpea flour in three litres of cold water with approximately 20 g of sea salt. The resulting mixture has to be deposed into a dedicated casserole and subsequently cooked until its 'boiling' point. It is essential that the mixture is continuously stirred during the process for a limited time (10 min). The aim of this procedure is to obtain a gelatinised structure by amylopectin and amylase. The original starch, composed mainly of amylopectin and amylose, is hydrated and consequently increases its resistance to movement, in the rheological sense. In this way, the initial pseudo-crystalline network is replaced with a more ordered gelatinised network with high viscosity. This behaviour is known as thixotropic phenomenon, roughly similarly to other non-food networks such as certain coatings (Djaković et al. 1990; Freundlich and Röder 1938; Parisi 2012). In this way, the original and difficultly digestible starch becomes a mixture of amylopectose and amylose chains with enhanced digestibility (provided that adequate enzymes are provided, such as in the human being).

The addition (20 g) of fresh parsley (or fennel seeds, as an alternative option) has to be performed after this step and before the complete recrystallisation of the amylopectose/amylose network. Anyway, the metastable mixture is deposed into wooden rectangular moulds (90 mm × 65 mm) with thickness not exceeding 3 mm. These moulds should allow the final (crude, before frying) *panella* slice to weigh between 15 and 20 g. After the recristallysation, *panella* slices may be detached and deposed into dedicated stainless trays until the final use. In addition, these forms may be stored in the fridge before frying.

The 'final' frying process has to be carried out at 180 °C and with seeds oil. It should be considered that there are no specific requests concerning peculiar oils: all possible 'seeds' oils are allowed. As a consequence (CNC 2014b), the table of ingredients should report 'various seeds oil' (a mixture of different seeds oils). It has to be considered that the nature of the crystallised amylopectin/amylose network allows *panella* slices to swell up. The browning or sautéing treatment (Sect. 2.3.2) should be carried out within one minute because of another reason: the yellow colour could be modified because of time excess. Moreover, there is a peculiar feature of the process: because of the re-crystallised network inside *panella* slices, two and more slices could adhere with possible ruptures. There are particular tools used by SF vendors: the use of pierced steel ladles with tools (*schiumarola*, in Sicilian dialect) allows the detachment and the concomitant removal of oil in excess because of the high number of pierces. It has to be considered that the ability of SF vendors appears critical in this step.

The obtained *panella* slices may be eaten 'as they are' or into bread sandwiches (Sect. 5.3.2.2).

5.3.2.2 *Pane e Panelle* Finishing

The obtained *panella* slices can be deposed into opened and partially deprived (of the inner content) *vastedda* breads. Sea salt or ground black pepper may be also used.

It has to be considered that, in accordance with the above-mentioned TSG proposal, the prepared *pane e panelle* SF should not be refrigerated or frozen (CNC 2014b). As above explained, the commercial option of refrigerated and under vacuum—thermosealed *panella* slices into supermarkets is tolerable for two reasons:

(1) The product is only *panella* instead of the final sandwich.
(2) Secondly, this product has to be fried.

5.4 Optional Notes and Differences

Some differences with the above-mentioned TSG proposal should be noted, on the basis of several references (Anonymous 2020b; Maranzano 2014; Zanni 2013):

(a) The use of salt for *pane e panelle* finishing may be associated with the addition of lemon juice. It should be considered that *pani câ meusa* can be prepared with lemon juice also (Chap. 4)
(b) An interesting but completely different version of *panella* slices may be found at the anniversary of St. Lucy in Palermo. These sweet *panella* foods are composed of chickpea flour with the addition of lard, table sugar, animal butter, and optionally *crème patissière*. Anyway, this is a different *panella* type: it cannot be used for *pane e panelle* SF.

5.5 Nutritional Profile of *Pane e Panelle* SF

The above-mentioned description of *pane e panelle* SF appears detailed enough when speaking of amounts concerning raw materials and ingredients. However, there are not many available data concerning the nutritional profile of this food.

We can use only some available references taking into account the following facts:

(1) The only available data are referred to the whole *pane e panelle* SF, but with a weight of only 200 g (while the single *vastedda* bread should be 200 g). This note has to be taken into account because of the important amount of water (100 g).

(2) Alternatively, there are some data concerning the single *panella* slices.
(3) There are not scientific papers concerning related durability in our knowledge. On the contrary, the above-mentioned TSG proposal states that *panelle* should not be stored under low temperatures. However, certain Palermo companies currently sell pre-packaged (thermosealed) crude *panella* slices to mass retailers and supermarkets.

Taking into account the above-mentioned premise, the following nutrition facts per 200 g should be assumed for *pane e panelle* SF (Anonymous 2020a, c; Basile 2015; Happyforks 2020):

– Energy (as kcal): 354 per 200 g),
– Lipids: 2 g,
– Saturated fat: 0.3 g,
– Carbohydrates: 75 g,
– Sugars: unavailable data,
– Fibres: 7 g,
– Protein: 13 g,
– Sodium: 300 mg,
– Cholesterol: 0 mg.

In addition, water is considered approximately equal to 100 g per 200 g of this SF. Actually, the *vastedda* bread is approximately 200 g. Anyway, some of these data appear questionable because the calculated energy intake (as kcal) on the basis of listed data (lipids, carbohydrates, and protein) is 370 instead of 354 kcal.

On a general level, the main amount of all expressed nutrients concerns surely carbohydrates; however, above-expressed doubts should not alter the general situation. The difference between calculated and stated energy intakes (16 kcal) can mean only four missing grams of protein or carbohydrates, or 1.8 g of 'hidden' lipids.

A note concerning these data is related to the declared sum of carbohydrates, lipids, and protein. This sum is exactly 90 g, while 100 g are the total dry matter (being water = 100 g). The remaining matter would be only 10 g. In this situation, more research should be needed in relation to this product. In fact, two references concerning single *panella* foods show clearly that lipids would be 11.04–11.44% on 100 g. On the other side, carbohydrates and protein would be 12.96–38.24 and 4.99–14.81%, respectively, and water would be approximately 68.9% (Anonymous 2020a, c; Happyforks 2020). These data should be considered only as part of the whole *pane e panelle* SF, but it is difficult to obtain a clear situation: the reliability of obtainable data should depend both on *vastedda* weight and the number of used *panella* slices. The only sure consideration is that SF contains mainly vegetable carbohydrates, protein, and oil; this fact mean has to be considered as the best nutritional expression of *pane e panelle* (vegetable meal) against 'animal' *pani câ meusa* SF. Moreover, some authors report that chickpea flour in *panella* slices contains a notable amount of polyunsaturated fatty acids… Once more, more research is needed (Buscemi et al. 2011).

References

Abulafia D (1978) Pisan commercial colonies and consulates in twelfth-century Sicily. Eng Historical Rev 93(366):68–81

Adjrah Y, Soncy K, Anani K, Blewussi K, Karou DS, Ameyapoh Y, de Souza C, Gbeassor M (2013) Socio-economic profile of street food vendors and microbiological quality of ready-to-eat salads in Lomé. Int Food Res J 20(1):65–70

Andolina C (2019) The Panormus Street food: la rosticceria palermitana conquista i mercati inglesi. www.orogastronomico.it. Available https://www.orogastronomico.it/news-ed-eventi/the-panormus-street-food-la-rosticceria-palermitana-in-inghilterra/. Accessed 7 April 2020

Anenberg E, Kung E (2015) Information technology and product variety in the city: the case of food trucks. J Urban Econ 90:60–78. https://doi.org/10.1016/j.jue.2015.09.006

Anonymous (2020a) Database degli alimenti e contacalorie - Panelle. www.fatsecret.it. Available https://www.fatsecret.it/calorie-nutrizione/generico/panelle. Accessed 7 April 2020

Anonymous (2020b) Panelle dolci. www.dolcisiciliani.net. Available https://www.dolcisiciliani.net/ricette/panelle-dolci/. Accessed 7 April 2020

Anonymous (2020c) Street Food - Cibo Di Strada - Palermo - Street Food - Cibo Di Strada - Palermo - Panino Con Le Panelle. www.myfitnesspal.com. Available https://www.myfitnesspal.com/food/calories/panino-con-le-panelle-657741024. Accessed 8 April 2020

Barbagli A, Barzini S (2010) Pane, pizze e focacce. Giunti Editore, Florence

Basile G (2015) Piaceri e misteri dello street food palermitano. Storia, aneddoti e sapori del cibo di strada più buono d'Europa. Dario Flaccovio Editore, Palermo. ISBN 978-88-579-0461-0

Buscemi S, Barile A, Maniaci V, Batsis JA, Mattina A, Verga S (2011) Characterization of street food consumption in Palermo: possible effects on health. Nutr J 10(1):119. https://doi.org/10.1186/1475-2891-10-119

Carrabino D (2017) 14 Oratories of the Compagnie of Palermo: sacred spaces of Rivalry. In: Bullen Presciutti D (ed) Space, place, and motion: locating confraternities in the Late Medieval and Early Modern City. Koninklijke Brill Nv (Brill), Leiden, pp 344–371. https://doi.org/10.1163/978900 4339521_016

Chukuezi CO (2010) Food safety and hyienic practices of street food vendors in Owerri, Nigeria. Stud Sociol Sci 1(1):50–57. https://doi.org/10.3968/j.sss.1923018420100101.005

CNC (2014a) Accordo tra Comune di Palermo e Consiglio Nazionale dei Chimici (CNC) per la partecipazione alla realizzazione di un sistema di salvaguardia e garanzia della tradizione gastronomica palermitana', Prot. 646/14/cnc/fta. Consiglio Nazionale dei Chimici (CNC), Rome. Available https://www.chimicifisici.it/wp-content/uploads/2018/10/20131210_accordo_firmato_dal_Presidente_del_CNC.pdf. Accessed 07 April 2020

CNC (2014b) DOMANDA DI REGISTRAZIONE DI UNA STG - Art. 8 - Regolamento (UE) n. 1151/2012 del Parlamento Europeo e del Consiglio del 21 novembre 2012 sui regimi di qualità dei prodotti agricoli e alimentary - "PANE E PANELLE". Annex to the document 'Accordo tra Comune di Palermo e Consiglio Nazionale dei Chimici (CNC) per la partecipazione alla realizzazione di un sistema di salvaguardia e garanzia della tradizione gastronomica palermitana', Prot. 646/14/cnc/fta. Consiglio Nazionale dei Chimici (CNC), Rome.Available https://www.chimicifisici.it/wp-content/uploads/2018/10/PANE_E_PANELLE__CNC_STG_2014.pdf. Accessed 7 April 2020

Crush J, Frayne B, McLachlan M (2011) Rapid urbanization and the nutrition transition in Southern Africa. Urb Food Secur Netw Series 7:1–49. Queen's University and African Food Security Urban Network (AFSUN), Kingston and Cape Town. Available https://fsnnetwork.org/sites/default/files/rapid_urbanization_and_the_nutrition.pdf. Accessed 7 April 2020

Dauverd C (2006) Genoese and catalans: trade diaspora in Early Modern Sicily. Mediterr Stud 15:42–61

Djaković LJ, Sovilj V, Milošević S (1990) Rheological behaviour of thixotropic starch and gelatin gels. Starch 42(10):380–385. https://doi.org/10.1002/star.19900421004

Duby G (1976) L'anno Mille: storia religiosa e psicologia collettiva. Einaudi, Torino

European Commission (2019) DOOR database. European Commission, Brussels. Available https://ec.europa.eu/agriculture/quality/door/list.html?recordStart=0&recordPerPage=10&recordEnd=10&sort.mileston. Accessed 7 April 2020

European Commission (2020) eAmbrosia—the EU geographical indications register. European Commission, Brussels. Available https://ec.europa.eu/info/food-farming-fisheries/food-safety-and-quality/certification/quality-labels/geographical-indications-register/. Accessed 7 April 2020

European Parliament and Council (2012) Regulation (EU) No 1151/2012 of the European Parliament and of the Council of 21 November 2012 on quality schemes for agricultural products and foodstuffs. OJ Eur Union L 343:1–29

FAO (1997) Street foods. FAO Food and Nutrition Paper N. 63. Report of an FAO Technical Meeting on Street Foods, Calcutta, India, 6–9 November 1995. Food and Agriculture Organization of the United Nations (FAO). Available https://www.fao.org/3/W4128T/W4128T00.htm/. Accessed 7 April 2020

Fellows P, Hilmi M (2012) Selling street and snack foods. FAO Diversification booklet number 18. Rural Infrastructure and Agro-Industries Division, Food and Agriculture Organization of the United Nations (FAO), Rome. Available https://www.fao.org/docrep/015/i2474e/i2474e00.pdf. Accessed 7 April 2020

Freundlich H, Röder HL (1938) Dilatancy and its relation to thixotropy. Trans Farad Soc 34:308–316

Guigoni A (2004) La cucina di strada Con una breve etnografia dello street food genovese. Mneme-Revista de Humanidades 4, 9 fev./mar. de 2004:32–43

Haddad MA, Abu-Romman S, Obeidat M, Iommi C, El-Qudah J, Al-Bakheti A, Awaisheh S, Jaradat DMM (2020a) Phenolics in Mediterranean and Middle East Important Fruits. J AOAC Int, in press

Haddad MA, Obeidat M, Al-Abbadi A, Shatnawi MA, Al-Shadaideh A, Al-Mazra'awi MI, Iommi C, Dmour H, Al-Khazaleh JM (2020b) Herbs and medicinal plants in Jordan. J AOAC Int, in press

Happyforks (2020) Panelle—sicilian chickpea fritters. Happyforks.com. Available https://happyforks.com/recipe/3692. Accessed 8 April 2020

Hawkes C, Harris J, Gillespie S (2017) Urbanization and the nutrition transition. Glob Food Pol Rep 4:34–41. https://doi.org/10.2499/9780896292529_04

Maranzano B (2014) Lo sviluppo del fenomeno "street food": il cibo di strada a Palermo ieri e oggi. Dissertation, University of Pisa, Italy

Ministero delle politiche agricole alimentari e forestali (2014) Quattordicesima revisione dell'elenco dei prodotti agroalimentari tradizionali. Ministero delle politiche agricole alimentari e forestali, Rome. Available https://www.politicheagricole.it/flex/cm/pages/ServeBLOB.php/L/IT/IDPagina/3276. Accessed 7 April 2020

Parisi S (2012) Food packaging and food alterations. Smithers Rapra Publishing, Shawbury

Parisi S (2019) Analysis of major phenolic compounds in foods and their health effects. J AOAC Int 102(5):1354–1355. https://doi.org/10.5740/jaoacint.19-0127

Parisi S (2020) Characterization of major phenolic compounds in selected foods by the technological and health promotion viewpoints. J AOAC Int, in press

Pisano A (2011) La farinata diventa fainé. Un esempio di indigenizzazione. Intrecci. Quaderni di antropologia cultural I(1):35–58. Associazione Culturale Demo Etno Antropologica, Sassari

Pitte JR (1997) Nascita e diffusione dei ristoranti. In: Flandrin JL, Montanari M (eds) Storiadell'alimentazione, Laterza, Roma and Bari

Scotto A (2010) La Panissa (profumi di storia savonese). www.trucioliavonesi.it. Available https://www.trucioliavonesi.it/index.php?option=com_content&view=article&id=4963:la-panissa-profumi-di-storia-savonese&catid=130:alessandro-scotto&Itemid=57. Accessed 7 April 2020

Sujatha T, Shatrugna V, Rao GN, Reddy GCK, Padmavathi KS, Vidyasagar P (1997) Street food: an important source of energy for the urban worker. Food Nutr Bull 18(4):1–5. https://doi.org/10.1177/156482659701800401

Tavella S (2019) Il panino con le fette di Savona. A tutta panissa (fritta). www.tastingthewo rld.it. Available https://www.tastingtheworld.it/panino-con-le-fette-savona/4635/. Accessed 7 April 2020

Vanschaik B, Tuttle JL (2014) Mobile food trucks: California EHS-Net study on risk factors and inspection challenges. J Environ Health 76(8):36–37. Gale Academic OneFile. Available https://go.gale.com/ps/anonymous?id=GALE%7CA365689844&sid=googleScholar&v=2. 1&it=r&linkaccess=abs&issn=00220892&p=AONE&sw=w. Accessed 7 April 2020

Zanni M (2013) Per Santa Lucia non possono mancare le panelle dolci: l'altra faccia delle "piastrelle" dorate e salate. www.sceltedigusto.it. Available https://www.sceltedigusto.it/public/ per-santa-lucia-non-possono-mancare-le-panelle-dolci-laltra-faccia-delle-qpiastrelleq-dorate-e-salate/. Accessed 7 April 2020